Y0-DTC-101

AN INTRODUCTION TO SEED TECHNOLOGY

An Introduction to SEED TECHNOLOGY

J. R. THOMSON
B.Sc. (Pure Science), B.Sc. (Agric.)

Course Director in Seed Technology
The Edinburgh School of Agriculture

Leonard Hill

Leonard Hill

A member of the Blackie Group
Bishopbriggs
Glasgow G64 2NZ

Furnival House
14–18 High Holborn
London WC1V 6BX

First published 1979

International Standard Book Number

0 249 44155 1

Printed in Great Britain by
Thomson Litho Ltd, East Kilbride, Scotland.

Preface

THIS IS NOT AN ENCYCLOPEDIA OF SEED PRODUCTION; NEITHER IS IT AN instruction manual for crop inspectors, a compendium of processing data or a laboratory handbook for seed analysts. It deals with the business of taking basic seed from the plant breeder and multiplying it so as to ensure to the farmer year after year a supply of high-quality seed. This requires the application of scientific knowledge from a wide spectrum of disciplines. The text describes in broad terms the various operations involved, seeking to develop understanding rather than detailed expertise. The aim is to help young people entering the seed industry, by bringing the different specialisms together into one text. It is set at post-secondary school level and assumes some knowledge of the basic sciences of botany and genetics.

I would like to acknowledge my debt to Dr James Delouche and his colleagues at the Seed Technology Laboratory of Mississippi State University. Information and ideas which have "spun-off" from their efforts to promote and guide the growth of seed improvement programmes in less-developed countries have been freely drawn upon.

For freedom to reproduce photographs, acknowledgements are due in particular to Agricultural Scientific Services of the Scottish Office and to the Seeds Division of Scottish Agricultural Industries Ltd.

J. R. T.

Contents

1

SEED QUALITY

SEED QUALITY IS A MULTIPLE CONCEPT COMPRISING SEVERAL components. In this chapter ten components are defined, but they are not all of equal value, nor is their order of relative importance the same in all circumstances. For a farmer, quality means suitability for sowing on his own farm, at a certain time of year, and for his own particular purpose. Seed health is more important in a wet than in a dry district, germination capacity in bad weather than in fine weather, seed size for mechanical rather than for hand planting, analytical purity for a main crop than for a catch crop, cultivar purity for malting barley than for stock feeding, and so on.

The farmer, his advisors and his seed merchant require an assessment of the quality of the seed before it is sown. Except for cultivar purity, laboratory methods of testing for the various attributes have been evolved. For the information of the farmer, test methods may vary to suit local conditions but, for commercial and legal purposes, uniformity of results is important and tests must be carried out by standarized methods. These should, however, give results which indicate the field sowing value, and such methods are described in the International Rules for Seed Testing.

Analytical purity

It is clearly desirable that the farmer should receive the species of seed that he wants. In most cases he is capable of looking after his own interests in this respect and can, for example, reject a delivery of wheat if he ordered barley. But it is not always as simple as that, and few farmers are capable of distinguishing meadow fescue from ryegrass seed, for example, or of recognizing the many impurities that may be present.

Analytical purity indicates how much of the material in the bag is intact seed of the species named on the label and is estimated by the analysis of a small sample in the laboratory. In the analysis, impurities are separated from the pure seed, i.e. from seeds of the named species. The impurities are seeds of other crop species, weed seeds and inert matter such as broken seeds, chaff, pieces of leaf, soil particles and so on. When the separation is complete, the pure seed is weighed and expressed as the *percentage by weight* of the whole sample.

Analytical purity is the basic component of seed quality, but it is not sufficient to state only the percentage figure—the nature of the impurities has to be taken into consideration. In grass seeds, the commonest impurities are empty florets—these are of no value, but they are innocuous, whereas weed seeds are positively harmful. A high purity percentage depends on the success of the cleaning operations after harvest. In most crop species percentages of 98 or more can be expected, but in some grasses empty florets are difficult to remove and the purity may be much lower.

Species purity

For some purposes the percentage analytical purity is not sufficiently precise. The percentage is expressed to the nearest 0.1, which is equivalent to the occurrence in the sample of two or three seeds of a size similar to the pure seeds. When barley is present as an impurity in wheat, for example, one or two seeds would be noted, but might not be reported as a separate percentage.

Furthermore, the sample analysed is so small that significant impurities may be missed altogether. In cereals, the sample is about one thousandth part of the quantity of seed sown on a hectare; so there could be about a thousand barley seeds in this quantity of wheat seed and yet none might appear in the analysed sample.

Consequently, when it is especially desirable to avoid contamination of one crop species by another of similar type, a larger sample is examined and the number of seeds of the other species is counted. The result is then expressed as the *number* of seeds in the weight of seed examined, e.g. two per kilogram.

Freedom from weeds

All farm land contains weed seeds. When a farmer buys seed, he will not willingly pay for more of them, but nevertheless a few seeds of the weed species already on his farm will do no serious harm. There are, however, some species of weeds which are not universally present on all farms and which, once established, are difficult to eradicate. Naturally, a farmer whose land is free from such weeds wants an assurance that there is none present in the seed he buys. Weeds of this kind are often described as noxious weeds; seed laws usually include special regulations about them, and certified seed should be free from them. The best-known example of such a seed is the wild oat.

Weed seeds are best expressed as the number found in the weight of seed examined, partly for the same reasons as seeds of other crop species are expressed in this way and partly for another reason. Weed seeds differ so much in size, that a percentage by weight is meaningless. For example, if the proportion of weed seeds by weight is 1 %, the number of weed seeds in a kilogram could be either 800 if the weed were *Ranunculus arvensis*, or 100 000 if the weed were *Cerastium vulgatum*. What is of importance is not the weight of weed seeds that is sown in a field, but the number.

Cultivar purity

Each of the common crop species includes a number of cultivars. These have either evolved by natural selection and are adapted to the regions where they are found, or have been deliberately produced by plant breeders. Cultivars differ in characters which are of practical interest to the farmer or to the user of the harvested product, such as yield, earliness, length of straw, water and temperature requirements, feeding value and suitability for manufacture. The seed bought by a farmer must be of a cultivar suitable for his farm, for sowing at a certain season, and for his own particular purpose. The problem for the seed technologist is to ensure that the seed is genuine and of high genetic purity.

As a general rule, cultivars cannot be identified by an examination of their seeds. It may be possible to say that the seed examined belongs to a certain group of cultivars, but it is seldom possible to identify the exact one. Cultivar purity therefore differs fundamentally from analytical purity in that it cannot be determined in a laboratory test.

A cultivar can be identified with more certainty by examination of growing plants. Cultivar certification schemes have therefore been set up in which the mother plants are examined in a general way on the farm of production, and a plot sown near the laboratory with a sample of the same seed is examined more meticulously. Cultivar purity determined in this way, however, cannot give the same assurance as a laboratory test carried out on a sample drawn from the actual bags supplied to the farmer, and it is supplemented by controlling the source of the seed sown for multiplication.

Germination capacity

High purity avails nothing if the seeds are incapable of germinating and producing strong seedlings in the field. The germination capacity of a lot is the percentage *by number* of pure seeds which produce normal seedlings in a laboratory test—seeds which are weak or abnormal in any way are ignored. This figure indicates the potential of a lot for establishing seedlings in good field conditions.

The importance of high germination capacity to the farmer is obvious, but in practice the full potential is seldom realized and the number of seedlings established in the field is less than the germination capacity would indicate. The conditions in the field at the time of sowing may be generally unfavourable—too dry, too wet or too cold—so that the seedlings die of drought or asphyxiation, or they are attacked by fungi. Even if conditions are in general favourable, individual seeds may fall on or under a stone, be attacked by soil insects, eaten by birds, or fail because of competition from weeds or other seedlings. A consequence of high mortality in the field is that small differences in percentage germination capacity are of no practical significance. Nevertheless, the higher the germination capacity, the better the field establishment, and in lots with high capacity the establishment approaches very close to it. This is well illustrated by the results shown in Table 1.1 of an experiment in which the laboratory germination of five lots of pea seed was compared with establishment in the field.

The germination capacity, in practice, is the best general indication we have of a seed lot's ability to grow in the field, and lots of low capacity should be rejected. As the figures in the table show, it is not possible to compensate for low germination by sowing proportionately more seeds. Germination capacity is influenced by harvesting and subsequent storage conditions, and the quality available is liable to vary from year to year.

In leguminous species, seeds occur which do not absorb water in a laboratory test, but remain hard and unswollen. These "hard seeds" appear frequently in forage crops, but less frequently in pulses, and are reported separately in the result of a germination test. They germinate later than normal seeds in the soil and indeed may lie buried for several years. Except in perennial forage crops, therefore, they have little value.

Analytical purity and germination capacity can be combined and expressed as one value, known as "pure live seed". This is calculated from the formula:

$$\frac{\text{(percentage analytical purity)} \times \text{(percentage germination capacity)}}{100}$$

It indicates how much percent by weight of the lot consists of seeds of the named species which are capable of germinating to produce robust seedlings.

Vigour

The germination capacity of a seed lot indicates its ability to establish seedlings in good field conditions; *vigour* indicates its ability to do so in poor conditions. The germination figure may therefore include seeds of insufficient vigour for good establishment on the farm. This is particularly the case in seed lots of low germination as shown in Table 1.1; in the poorest lot of that experiment only one third of the germinable seeds established seedlings in soil. High germination capacity, on the other hand, was associated with high vigour, and this is the general rule, but it does not follow that seed lots with the same high germination capacity have equal vigour.

Table 1.1 Germination capacity and field establishment of five lots of pea seed (from Franck, W. J., *Proc. Int. Seed Test. Ass.*, Nos. 9/10, p. 1, 1929)

Percentage of seeds producing seedlings in the laboratory	Percentage of seeds producing established seedlings in the field		
98	91	= 93%	of laboratory germination
93	68	= 73%	
80	39	= 49%	
71	33	= 46%	
56	16	= 29%	

Because there are many agronomic factors which can be inimical to seedling establishment, such as soil moisture, texture, temperature and micro-organisms, there may be correspondingly different kinds of vigour. Vigour has no precise physiological definition, but nevertheless some very useful vigour tests have been developed for certain species, e.g. the cold test for maize and the conductivity test for peas. These can be applied when it is known that the seed is to be sown in sub-optimum conditions.

Vigour can be affected by damage to the embryo incurred during harvesting or subsequent processing. Seeds of pulses are particularly susceptible to such damage, but the effects can often be observed in a standard germination test, and the worst cases evaluated as abnormal seedlings. The testa, or other seed covering, provides protection against physical damage and the entry of parasites and, by checking the diffusion of gases, can restrict respiration and so prolong storage life. Seeds with intact seed coats are therefore more likely to retain their vigour during post-harvest operations, storage and planting.

Other factors affecting vigour are environment and nutrition of the mother plant, stage of maturity at harvest, seed size, senescence and pathogens. Low vigour is sometimes associated with changes in the permeability of the cell membranes arising from either immaturity at harvest or poor storage conditions. Normally the cell membranes are semi-permeable, admitting water but preventing the outward movement of dissolved substances when the seed is immersed in water. This semi-permeability may be lost or modified, so that soluble nutrients diffuse out of the seed and encourage the growth of micro-organisms in the soil, and so influence seedling establishment.

Seedling vigour is also a genetic character, as shown most spectacularly in hybrid cultivars. Discrepancies between germination capacity and field performance are more common in some crop species than in others—in pulses, for example, in comparison with cereals—and there are similar differences between cultivars of the same species.

Size

This quality has two components: actual size and uniformity of size.

Any seed lot, as harvested, includes seeds of different sizes. This variation is partly due to differences between the seeds harvested from different plants and partly due to differences between seeds borne on the same plant.

The variation between plants is due to genetic and environmental factors. Seed size is to some extent an inherited character, and genetic variation is more likely to occur in cross-pollinated than in self-pollinated species. Seed size is also influenced by the nutrients, minerals and moisture available to the developing seed, and so depends on factors which vary from plant to plant throughout the crop—shading, soil moisture and fertilizers—and which are modified by local competition. Seed development can also be retarded by diseases and pests which may not attack every plant.

The size of seeds borne on the same plant depends on their position on the plant within the inflorescence. Seeds developing on the lower shaded parts tend to be smaller than those borne close to leaves exposed to full daylight and, within an inflorescence, the largest seeds are near the base. Flowers on lateral branches open later than on the main stem and give rise to seeds which are liable to be harvested while still immature and undersized.

Large seed size is an indication of vigour. The seed contains either an embryo alone, or an embryo plus endosperm ready to be absorbed into the developing seedling. In either case, the bigger the seed, the greater the size of the seedling and the area of green leaf capable of photosynthesis. If seed is buried deep in the soil, a large seedling is more likely to reach the surface than a small one. Compared with small seeds, therefore, large seeds produce seedlings which grow more rapidly in the field, and a greater proportion of them emerge through the soil surface. Indeed, small shrivelled seeds have no practical planting value. It does not follow that a bigger yield is always obtained by sowing large seeds. When plant populations are similar and field conditions are good, plants from small seeds have grown sufficiently by harvest time to produce an equivalent yield of grain.

The effect of size is most obvious in comparisons between different species. When a cereal crop is sown, at least 50% (and usually more than 80%) of the germinable seeds may be expected to produce established seedlings. At the other extreme, small seeded grasses may achieve an establishment of no more than 20%.

Seed size is usually expressed as the weight of a thousand seeds. Alternatively, though less precisely, it may be expressed as the weight of seeds that can be contained in a certain volume, such as a hectolitre.

Uniformity of size is of importance for several reasons. It can influence the effectiveness of seed cleaning operations. For example, in clovers, large seeds are more severely abraded in the hulling machine than are small seeds, and this has two effects on the larger seeds, one beneficial and one harmful– making hard seeds permeable and damaging the embryo. More important, uniform size makes for uniform growth of the seedlings, so that the growth of a plant is not retarded by the shading effect of a larger neighbour. It enables a mechanical drill to distribute the seed more evenly in space and depth than is possible with a mixture of large and small seeds. It is of most significance, however, for crops in which the seeds have to be spaced widely apart along a row as in maize. When planting is done mechanically, the dimensions of each seed may have to be approximately the same in order to fit into the notches or slots of the machine.

In maize there are various sizes of kernel on each ear. The smallest are rejected solely because of their size. The largest are not acceptable for mechanical planting, because there are not enough of them of uniform size; they can, however, be sold in small lots for hand planting. Seed for mechanical planting is taken from the large quantity of intermediate size. For the conventional plate-type machine, the seeds must not only be uniform in size (i.e. in volume and weight) but also uniform in shape (i.e. in their dimensions), and strict grading is essential—but a new plateless type of planter is now being introduced which does not require such close uniformity.

A further development along the line of fitting the seed to the planting mechanism is the seed pellet—a structure containing a single seed, the seed being surrounded by inert material so as to form a regular shape, usually spherical, of uniform size. Pelleted seed is used for precision sowing of spaced crops such as sugar beet and vegetables. It is mechanically impossible for seeds of certain species to be placed at regular intervals along a row because of irregularity in size and shape. The traditional method is to sow a continuous line of seeds along the row and thin out the superfluous seedlings, leaving widely spaced single plants. Pelleted seeds can be placed at exact spacings by machine. The seeds may be sown to a stand, i.e. one seed placed for each plant required; ideally this requires 100 percent establishment, but a plant population with some gaps may be acceptable. If gaps are not acceptable, or if the germination capacity of the seed is not high enough, the seed may be placed at, say, half the required spacing and the superfluous seedlings removed. The labour requirement for this operation is much less than for conventional thinning.

This treatment may also be used in cases where it is required to spread a limited amount of seed over a large area. For example, when a small amount of breeder's seed of clover or lucerne has to be multiplied to the greatest possible extent by establishing widely spaced plants.

Other possibilities are granulates and tapes. Granulates are made to facilitate sowing of very small seeds such as *Agrostis*; each granule contains up to three or four seeds. Tapes are made for vegetable growing; the tape is a collapsed tube made of a water-soluble plastic, containing single seeds spaced along its length. When it is buried in a shallow furrow, the plastic material dissolves and allows the seed to germinate in position.

Uniformity

Every seed lot is to some extent a mixture—of pure seed, inert matter, crop seeds and weed seeds, and of live seeds and dead seeds. A lot should, therefore, be thoroughly blended before it is packaged, so that the contents of every bag are the same. Unfortunately, much of the seed sold to farmers fails to reach a high standard of uniformity. This is recognized in the warning printed on international certificates, that the analysis results apply to the lot as a whole and not necessarily to any part of it. This situation has arisen partly because simple but effective methods of blending have not been available, and partly because the test for uniformity involves a great deal of sampling and testing work. It may not be practicable to apply the test to every lot, but at least it should be used to assess the effectiveness of any new blending equipment or procedure that is introduced.

Health

Seed health is important in controlling certain crop diseases and in ensuring good field establishment. Parasitic organisms causing disease must have some means of surviving the dead season between crops, and in some cases the infection is carried over from one year to the next by the seeds. Such diseases are said to be *seed-borne*.

In one type of seed-borne disease, the seedling from an infected seed grows vigorously, but at a later stage in the crop's development spores are produced which can spread the infection throughout the crop. Good examples are the smut and bunt diseases of cereals, which do not produce spores or any symptoms of disease until the flowering stage of the plant is reached.

Another type of seed-borne disease destroys or stunts the growth of the seedlings and rapidly spreads to neighbouring plants, causing poor field establishment. The survivors produce infected seeds which, in their turn, give rise to infected seedlings in the following year. Examples are the many leaf-stripe and root-rot diseases of cereals. The harmful effects of such diseases on field establishment are not always indicated by the result of a standard laboratory germination test.

Certain viruses are carried within the embryos of infected seeds and cause mosaics or leaf-stripe diseases on mature plants. In the field, infection is commonly spread from diseased to healthy plants by insect vectors; rarely, it may be transmitted by pollen.

Seed-borne diseases can be controlled by chemical treatment of the seed, but this requires the widespread use of chemicals which may be poisonous to man and animals. A better policy is to sow seeds harvested from healthy crops and, to achieve this, some seed certification schemes impose standards of health. These standards may be based on the incidence of disease in the crop from which the seed is harvested and in neighbouring crops, or on laboratory tests of the seed to be certified. These tests identify the pathogens present and estimate the percentage of seeds infected. Tests can also be performed to check that any chemical treatment has been carried out effectively.

Apart from micro-organisms which cause disease, seeds carry saprophytic fungi and bacteria on their surfaces and in the superficial layers. Normally these are harmless and do not invade the embryo unless it has been injured.

Moisture content

Moisture content is of interest to the processor and the store manager rather than to the farmer. It is the key factor in determining whether or not seed will retain its germination from harvest to sowing time. It can be measured in a laboratory or, more quickly though less accurately, on the spot by moisture meters.

Conclusion

It is the task of the seed technologist to take seed of a cultivar from the plant breeder every year and multiply it to produce sufficient quantities of high-quality seed for sale to farmers.

To sum up, high-quality seed should be:

of high analytical, species and cultivar purity
free from weed seeds
of high germination capacity and vigour
of uniformly large size
free from seed-borne diseases
of low moisture content

Cultivar purity is a measure of the seed's genetic quality as created by the breeder, and the other attributes determine the extent to which this genetic potential can be realized on the farm. High germination capacity and vigour enable the seed rate to be reduced. This compensates for the additional cost that is incurred in the production of high-quality seed and reduces the amount of harvested produce that has to be diverted from the nation's food supply to provide the following year's seed.

Historical development

Consistent improvement in seed quality dates from 1869, when Friedrich Nobbe opened a seed testing station in the small town of Tharandt, situated in what is now East Germany.

In primitive farming in Europe only cereals and pulses were grown. The seed was harvested; most of it was eaten, but some was held over until the next year for sowing. These crop species have large seeds which are easily recognizable; the seed that was sown may have contained a great deal of impurity, but at least the farmer knew that he was sowing the right species. Low germination due to a bad harvest or poor storage conditions would often be associated with discoloration, which would serve as a warning. There must, nevertheless, have been many crop failures due to poor-quality seed, but this was probably not appreciated at the time.

About 300 years ago new crops began to be introduced, such as turnip, clovers and grasses, and these were not grown for their seeds but for their vegetative parts. Farmers made hay from natural pastures, and this was stored in barns to feed livestock during the winter. This hay bore ripe seeds which became detached and accumulated on the floor of the barn. The first herbage seed to be sown by farmers was obtained in this way, and inevitably was a mixture of grasses and clovers, useful species and weeds. From this developed the deliberate practice of cutting hay in order to obtain seed from it. One result of this was that seed production became separated from crop production; it tended to become a specialized business concentrated in favourable areas. Some farmers had seed to sell, others had need of it, and so the seed trade began. Another result was to emphasize that there was no way of assessing seed quality. Seeds of the new crops were small and difficult to distinguish, even if their species were known, and the contaminating weed seeds could not be recognized. Farmers could not distinguish between ryegrass and fescue, between turnip and charlock, or even between white clover seed and grains of sand. Viability was a mystery.

With the agricultural revolution in Europe came the realization of the need for some method of assessing the value of seed before sowing, but an understanding of seeds had to await the development of botanical science. It was not until last century that the seeds of different species could be identified and enough knowledge was acquired about seed structure and the physiology of germination. When quality was defined and made measureable, it became possible to improve it.

By 1869 knowledge had advanced far enough for Professor Nobbe to make his laboratory available for testing agricultural seeds. Though the stimulus had come from the new crops, the traditional crops also benefited from the new knowledge. Both seedsmen and farmers had their seeds tested for analytical purity and germination, and guarantees of quality were given accordingly. This initiative was rapidly taken up and, within 50 years, laboratories providing a testing service had been established all over Europe and North America. Today there are official seed testing stations in about 60 different countries, prepared to test agricultural seed of more than 200 different species for nine different attributes. Seed laws ensure that seed sold to farmers is of high quality, and tests for enforcement and advisory purposes are carried out at these stations.

An early development was the organization in some countries of certification schemes. In these schemes, standards were set for quality attributes that could be assessed by laboratory tests, and seed lots which attained these standards were sealed and had distinctive labels attached. Such schemes were managed officially, but participation was voluntary. Certified seed was available for purchase by farmers who wanted it, and who appreciated the assurance of high quality.

The availability of a testing service encouraged the seed trade by making it possible to specify the quality of seed offered for sale and to confirm that deliveries were in accordance with the specifications. Trade across international frontiers, however, was hindered by lack of coordination between the seed testing stations. Because of differences in the definitions, methods, equipment and materials employed, there were serious discrepanceis between the results of tests of the same seed lot performed by laboratories situated in different countires.

The need for uniformity in testing was met by the formation of the International Seed Testing Association (ISTA) in 1924 for the purpose of discovering the causes of these discrepancies and eliminating them by devising procedures and techniques which could be followed by every seed testing station in the world. The association pursued this aim by organizing comparative and experimental testing, and by bringing specialists from different countries together for detailed discussions.

In comparative testing, samples from the same seed lot are sent to all member laboratories for test, and the test results are subsequently compared. This indicates whether or not divergence between stations is excessive. It also shows each participating laboratory how it compares with other laboratories, and may draw attention to unsuspected faults in test procedures. The purpose of experimental testing is to compare and evaluate different methods. A limited number of stations participate in testing samples from several seed lots of the same species with sufficient replicaion for statistical analysis of the results.

The result of a purity test depends to a large extent on the definition of pure seed; the definition used in America differed from that used in Europe and, in the case of grasses, there was another one in Ireland. Similarly, the result of a germination test depends on the concept of germination; some laboratories regarded the seed as having germinated if it merely showed evidence, by the protrusion of the radicle, of viability, while other laboratories related the test to field planting value, and did not regard a seed as having germinated unless the seedling was robust enough to survive in field conditions. It was not until 1953 that universally acceptable definitions of purity and germination were negotiated.

Agreed sampling and testing methods are published by the association in the form of rules. Quality standards are not laid down, but only methods by which quality is to be measured. That a seed lot has been sampled and tested by these methods is indicated by issuing the test report on a special international certificate, and this may be done only by authorized stations. The rules are amended from time to time to incorporate new scientific discoveries, or to provide for changes in farming practice, such as the introduction of pelleted seed.

For seed moving in international trade, sampling and testing in accordance with the rules is, by custom if not by law, obligatory. In testing for domestic purposes, no such obligation exists, but the methods described are based on scientific evidence and long experience, and should be followed unless there are strong local reasons for modifying them. If the international methods are not followed, imported seed on arrival will be tested by methods which are not the same as in the exporting country, and troublesome discrepancies will arise. Different methods can be justified in response to any peculiarity of locally produced seed, such as the nature of the common impurities or the type of dormancy.

During the last century, distinct cultivars were selected and propagated by individual seedsmen and farmers, and seed of these cultivars (though sometimes of doubtful authenticity) was available for sale. Early this century, in some countries, voluntary associations of breeders, merchants and farmers were established to organize and control the multiplication of such cultivars, and thus assure the genuineness of seed sold to farmers.

While Friedrich Nobbe was working in Tharandt, Gregor Mendel was studying the inheritance of plant characters at Brunn in Austria. It was not until 50 or 60 years later, however, that his discoveries were applied to the production of new cultivars. The effect of these new cultivars has been spectacular. The wide adoption in the corn belt of hybrid varieties betwen 1938 and 1945 was associated with an increase of 15–20 percent in the average USA yield. Other factors have been involved in such productivity increases, e.g. fertilizers, herbicides, seed treatment and mechanization, but in Sweden it is reckoned that new cultivars contributed 25 percent of the increased yield per hectare of wheat in the first half of the century and 12 percent in other cereals. In England and Wales the yield of cereals increased by 40 percent in the 20 years up to 1960, and half of this is thought to be attributable to new cultivars.

With the release of this stream of new cultivars to farmers, official control of multiplication became essential, and schemes for the certification of cultivar purity were set up. Cultivar certification cannot be based on a test of the seed that is to be sold, but involves some control over production, and so is more complex than previous schemes which certified laboratory standards only. These original schemes have become obsolete, and certification now covers cultivar purity at least.

By the 1950s it was realized that the proliferation of national cultivar certification schemes with different nomenclatures and control systems was impeding the free flow of good seed across international frontiers, and a plan for harmonization was evolved by the Organization for Economic Cooperation and Development (OECD). This organization was set up in Paris to promote the post-war economic recovery of European countries, but its seed scheme is open to all member countries of the United Nations. The scheme was launched in 1958: it now has more than 30 participating countries and provides for all the major crop species. It does not fix standards of seed quality, but defines the procedures to be followed and the grade names to be used. A seed lot which has been certified in accordance with the OECD's rules and directions is entitled to a distinctive label, which is recognized by the import control authorities of participating countries. The scheme includes a system for controlling the multiplication of seed in favourable climatic conditions outside the country of origin of the cultivar.

The Food and Agriculture Organization (FAO) of the United Nations was established in 1948, principally for the purpose of increasing agricultural production in the underdeveloped parts of the world. From the beginning the Organization has realized that one of the most effective means of achieving this aim is to promote the improvement of seed quality. Accordingly, it has given technical aid and encouragement in such ways as organizing training courses, encouraging plant breeding, coordinating international seed health requirements, recommending seed certification standards, and sending out technical missions. Its main effort in this field is now directed towards the development of seed industries.

FAO, OECD and ISTA are governmental bodies, but most of the world's seed industry is managed by private enterprise. In the international field, the interests of the private sector are watched over by the Fédération Internationale du Commerce des Semences (FIS). This organization was established originally for the purpose of facilitating the international movement of seeds by preparing rules and usages for the conduct of trade, and it has a long record of cooperation with ISTA in promoting the acceptance of the

Association's testing rules. The Fédération participated in the planning of OECD's certification scheme, and in the initiation of seed industries in developing countries is consulted by FAO.

2

THE SEED INDUSTRY

THE SEED INDUSTRY COMPRISES ALL THE COMPLEX INTERLOCKING operations that are necessary to ensure a regular supply of uniformly high-quality seed to farmers. It is one of a group of industries which together make up the great primary industry of agriculture. It cannot be studied in isolation and, before analysing and examining the different operations involved, its relationship to the general agricultural situation and to the social, economic and political structure within which it must function, needs to be considered.

Virtually the whole of primitive man's life was devoted to securing his food supply, at first by hunting and by collecting the edible parts of wild plants, and later by planting and harvesting crops. As farming methods improved, it became possible for a man to produce more food than was necessary for himself and his family. From this emerged a social structure in which, at its simplest, the community was divided into farmers and others, the others obtaining their food from the farmers by barter for goods and services, or by purchase.

In some countries the productivity of agriculture has increased to such an extent, particularly within the last century, that today less than 10 percent of the population are engaged in food production. In other countries, development has proceeded much more slowly, and about 60 percent of the population are still dependent on farming for a living. These less-well-developed countries have now to cope with a situation which did not arise in the history of the more-developed countries—the explosive expansion of the population due to application of biological science, particularly in the medical field. In European countries, for example, population growth was more gradual.

In many countries, therefore, a rapid increase in agricultural productivity is necessary to meet the ever-increasing demand for food. New agricultural policies are being evolved in general terms, and plans are being worked out in detail. To get more out of agriculture, it is necessary to put more in. The necessary inputs are partly materials and partly services. Materials include fertilizers, pesticides, irrigation water, machinery and energy; services include transport systems, research to improve farming methods, extension work, technical training, marketing organizations and financial credit. The critical input upon which all the others depend for their full effectiveness, is, however, high-quality seed, and in particular its genetic quality, i.e. the seed must be of cultivars capable of giving an economic return for such inputs as fertilizers and irrigation.

The overall plan must be kept in balance. Irrigation projects make heavy demands on a country's capital resources; the available energy supply may limit the manufacture of fertilizers, and import of chemicals may strain the balance of trade. It is possible to express crop yield in terms other than "tons per hectare," e.g. as kilograms per cubic metre of irrigation water, per kilogram of nitrogen, or per kilogram of pesticide. The aim is not maximum yield at any cost, but a cropping system that makes the most efficient use of inputs. The ideal cultivar may be quick-maturing, have a high nitrogen response or good disease resistance, thus economizing in irrigation water, fertilizer or chemicals. Of the various inputs, seed is the only one that is produced by farmers for farmers. It does not require vast sums of capital, and it does not have to be imported, but it does require local effort and initiative, and much detailed planning.

To provide the inputs, some transfer of population from agriculture to manufacturing and service industries is necessary. The farmer must be given an incentive to improve his productivity, and this requires both marketing arrangements to ensure that local or national surpluses do not depress the price of his produce, and the manufacture of consumer and luxury goods on which he can spend his additional income. The non-farming sector of the population must be prosperous enough to buy the food he produces. All this involves a considerable social upheaval and the creation of an economy that is prosperous in all sectors. Inevitably there will be a reduction in the farming population; the political problem is how to ensure that people so displaced find other employment.

Reducing the number of people employed in agriculture may not necessitate removing them from the countryside. Growth of towns involves huge expenditure on social servies such as sewage disposal, water supplies and housing. It may be, therefore, that industry should be organized in numerous small factories spread over the country, with the emphasis on manual labour rather than on machines.

The seed industry exists to serve the needs of farmers, and it has developed to varying levels in different countries, keeping in step with agricultural development as a whole. It is a specialized undertaking, operated by men who are experienced, knowledgeable and specifically trained for it. Adaptability and initiative, combined with a high level of managerial, financial and technical skill, are necessary in senior positions.

In Western Europe and North America, for example, it is mainly operated by commercial companies owned by men who have contributed the capital and who have a financial interest in its success, but with government control and some governmental participation. Farmers' co-operatives have also played an important part. In these countries the present position has evolved in response to the changing needs of agriculture; at no stage did a group of experts sit down and plan a seeds industry for the future. In other countries, such as in eastern Europe, the seed industry is almost entirely managed by government and the organization is to a large extent the result of deliberate planning.

Plans now being made for agricultural growth in under-developed countries must of necessity include proposals for the establishment of a modern seed industry. The experience of other countries suggests that the industry need not be entirely governmental nor operated entirely by private enterprise, but that a mixture of both is perhaps best.

Commercial companies have been reluctant to participate, except in the production of seeds of vegetables and F1 hybrid cultivars, which farmers cannot produce themselves and for which there is consequently an annual demand. In Europe and North America, seed companies started as small local concerns and gradually expanded, utilizing their own experience and re-investing their profits. The plans prepared for developing countries, however, involve large-scale operations right from the start; there is little practical experience on which to build, indigenous expertise is limited, the crop species may be strange, and the climate unfamiliar. There may also be uncertainties about the transfer of profits and the repatriation of capital. The risks are therefore high and the profits are not considered commensurate.

Large-scale commercial participation is therefore the exception, and the initiative and the capital investment normally come through government. It is realized, however, that, apart from testing and quality control, the most effective kind of organization is one modelled on a commercial company rather than on a government department. This allows the flexibility that is needed in operations which are seldom completely predictable, as well as promoting efficiency and cost control.

An important function of a national seed company is to carry over surplus stocks from one year to the next, and to build up and maintain reserve stocks in case of natural disasters such as drought and flooding.

The functions to be performed by a seed industry can be set out as follows:

Plant breeding—including genetic research
Cultivar assessment
Multiplication, i.e. growing seed crops on farms
Processing—including drying, storage and packaging
Marketing and procurement
Control—including legislation, certification and testing
Quarantine
Extension activities

Plant breeding, cultivar assessment, quarantine and extension activities are special subjects of study and may not require "seed technologists" in the strict sense. Nevertheless, they are essential parts of a seed industry, and practising seed technologists should have a general understanding of them.

Plant breeding

Local cultivars of crop species are acceptable in quality to the native population and are in many ways well adapted to the conditions in which they are grown, but they are incapable of giving yields that would justify the expenditure of money on the various inputs mentioned above. A policy of agricultural development cannot become effective simultaneously over the whole country; some regions will lag behind others, or may even retain traditional methods indefinitely—because of climate, the nature of the soil or because of sheer remoteness. Even in districts where conditions are favourable to high yields, some farmers are reluctant to adopt new methods of husbandry. It is therefore necessary for the plant breeder to produce new cultivars of two types. Firstly there is a need for a range of cultivars which, while maintaining an acceptable quality, are capable of high yields with heavy fertilizer applications, improved husbandry and, if available, irrigation water. Secondly, there is a need for cultivars which can give superior yields under traditional farming methods through improvement in such characters as early maturity, drought tolerance and disease resistance. These are sometimes described as "high technology" and "low technology" cultivars respectively.

Plant breeding is a slow, painstaking business and twenty years is a realistic period for a breeding programme to achieve its objectives. Nevertheless, in the meantime *some* improvement can be achieved in a few years by selection from traditional cultivars and the screening of improved cultivars grown in neighbouring countries. In practice, plant breeding is never finished. A new cultivar may be launched because it is superior in certain respects, though in other respects it is still capable of improvement; resistance to new diseases or new strains of existing diseases may become necessary; changes in farming practice may require cultivars suitably adapted. Furthermore, developments in the science of genetics may enable breeders to introduce novel characteristics. For this reason the interest of breeders should not be restricted to the immediate production of new cultivars. To ensure long-term progress, it is necessary to have some scientists engaged in fundamental research in genetics and related biological fields.

A plant breeding station has more than a creative function; it is not sufficient to produce a new cultivar as an isolated act of innovation and then lose interest in it. There is thereafter the essential function of maintaining the cultivar in a genetically unchanged state, and issuing small quantities of seed every year for further multiplication and eventual issue to farmers.

Plant breeding stations may be government institutions or may be operated by commercial seed companies. In developed countries both types co-exist. Under a planned agricultural economy, breeding usually starts as an official activity, but there is no reason why commercial enterprise should not participate at a later stage. Basic research, however, is likely to be restricted to universities and government research institutes, the results being made available to all through scientific journals.

Cultivar assessment

In order to assess the agricultural value of new cultivars, a plant breeding station has to carry out field trials. On the evidence of these trials a decision can be made as to which, if any, of several promising novelties are worth further multiplication. In addition, a nationwide system of performance trials is necessary in which new cultivars are grown alongside established cultivars and compared with them in yield per hectare and other agronomic features such as maturity, straw strength and disease resistance. To this organized trial system, plant breeders can submit the best of their new cultivars for evaluation. This national system needs to be organized and controlled by an independent authority set up for the purpose by the government. Its judgments must be absolutely unbiased and recognized as such.

To provide this service, the authority operates trial centres distributed throughout the country. A centre may operate on a government farm or on part of a privately owned farm which has been brought completely under the control of the authority for the duration of the growing season. At each centre, submitted cultivars and control cultivars are grown in replicated plots laid out in accordance with biometric principles. Observations and measurements are made throughout the growing season, and finally the produce of each plot is harvested and weighed (see chapter 13).

Before approving a new cultivar, the authority has to satisfy itself that it is indeed new and not an old one masquerading under another name, and that it is uniform and stable. Determination of distinctness involves a meticulous examination of the morphological details of individual plants and a study of their growth habits and rhythms. On this basis a botanical description is compiled, which is subsequently used in certifying seed as authentic.

A cultivar is a population of plants which have common ancestors and which have certain characteristics in common. Absolute identity of all the individuals in the cultivar may not be necessary, but a fair degree of uniformity is required for two reasons: the farmer expects his crop to be uniform in the characteristics for which it was bred, such as yield or drought resistance, and for certification purposes a cultivar cannot be identified if it is a mixture of types.

Genetic stability is necessary if a cultivar is to reproduce itself from seed and remain reasonably true to type over a number of generations.

Multiplication

Every year the breeder of a cultivar issues a small quantity of authentic seed, and this has to be multiplied over a number of generations to produce the quantity required for sale to farmers. In the case of a cereal cultivar, if the breeder issues enough seed to sow one hectare, for example, two or more generations will be necessary to produce enough seed to sow 1000 hectares. The first two or three generations are produced under the control of the plant breeding station. This may require quite a large area, which for state-bred cultivars can be provided on experimental or other government-owned farms.

Under a controlled system of multiplication, later generations can be produced on approved private farms under a certification scheme. The farmers receive their seed either through a cooperative association or a commercial firm, and dispose of their harvested produce through the same channel at a premium price of, say, 15–25 percent above the current grain price. This operation can involve a combination of official, commercial, cooperative and private agencies. The control that is necessary to prevent contamination can be exercised without much difficulty on government and other large farms, but becomes less effective when the seed is distributed to numerous farms and multiplied on small areas. Supervision of these private farmers is the function of the crop inspectors of the official certification authority, with the cooperation of the extension service and of any commercial companies involved.

Processing

The harvested seed goes from the farms to factory-like processing plants located in the seed-producing districts. There the seed is dried, cleaned, stored, treated and packaged ready for sale the following year.

In choosing sites for these processing plants, several factors have to be taken into consideration. A plant should be located near to the producing farms rather than to the ultimate seed purchasers. If drying is required, seed must reach the drier as quickly as possible, and long-distance transport is cheaper and safer after the seed has been cleaned and properly packaged. At the same time, for efficient working, the plant should be accessible to big enough quantities of newly harvested seed from the farms to ensure that its capacity is fully utilized. Other factors are the availability of power and labour, and the configuration of the road system.

Marketing

From the warehouses of the seed companies the seed is distributed to towns and large villages throughout the country for sale to farmers. For this purpose any type of retail outlet can be utilized provided it has sufficient safe storage and an interested manager. It may be company-owned, a co-operative store, or a privately owned shop, selling not only seed, but fertilizers, pesticides and other farm requisites. Experience has shown that distribution of seed to farmers is not a suitable function for the extension service. Too much time is taken up by store-keeping and accounting activities, and good extension workers are diverted from more important work.

High-quality seed is wasted unless it is indeed taken up for sowing, and the retail system needs to be managed with commercial vigour. A widespread transport system is necessary, and this may depend on the road and rail infrastructure set up under a wider development programme for agriculture as a whole. In other words, distribution and sale of seed should be co-ordinated with the marketing of inputs and outputs in general.

No country is completely self-sufficient in seed. Policy should aim at home production of all seed required for the major crops, but in years of shortage it may be necessary to import some. Seeds of minor crops might have to be imported because suitable conditions do not exist at home. Such imports are most effectively organized through the international network of commercial firms, but it is a function of a national seed company to ensure that seeds of all kinds used by farmers are in fact available in local markets.

Important elements in marketing are continuity and a reputation for reliability, not only in the performance of cultivars but also in their availability. When a farmer has taken up a new high-yielding cultivar and found that it suits him and his farm, he is naturally reluctant to change to another. In making the initial change from a traditional cultivar, he believed that seed of the new one would continue to be available; he may be persuaded to change again, but frequent changes antagonize him. For this reason, when a new cultivar is introduced, the one which it is expected to replace should not be withdrawn immediately.

Control

In order to ensure that the seed sold to farmers is indeed of high quality, it is necessary to exercise some control over the industry. The control must be objective and free from influence by any commercial or personal interest, and it is obviously the function of government or of some authority officially designated for the purpose.

Control involves the monitoring of seed quality throughout the various stages of production, processing and marketing. In a previous chapter, ten components of quality were described and all but one (cultivar purity) can be determined by laboratory tests. The first essential for control is, therefore, an official seed testing service, partly for enforcement purposes, and partly to provide an advisory service to all sectors of the industry. There may be one large seed-testing station for the whole country, or a number of smaller stations located in seed-producing districts; but in the latter type of organization it is essential that there be unified technical control to ensure that all the laboratories conform to the prescribed testing methods. The ISTA rules for testing provide a model, but for internal purposes need not be adopted in their entirety. In addition to official stations, there may be small private laboratories located in major processing plants, which provide the management with prompt tests of the material being processed and thus promote efficiency. Although tests are available for nine attributes, tests for all of them are not necessarily made on every sample submitted to a laboratory. The tests most frequently carried out are for analytical purity, weed content, germination capacity and moisture content.

The one attribute that cannot be controlled through laboratory tests is cultivar purity. This requires some degree of control from the growing crop right through to marketing of the harvested and processed seed. For this purpose a certification authority is necessary, with a corps of inspectors engaged in visiting farms, processing plants and retail outlets. For efficiency in the use of staff, laboratories and land, this authority should work in association with the seed-testing service and the cultivar trials authority, and be responsible for the monitoring of seed quality in general.

In order to control the industry, the government may need special powers, and the law which confers and defines these powers is usually known as the Seeds Act. This legislation is an essential step in the development of a seed industry, but it should be realistic and not set up restrictions which cannot be enforced (see chapter 18).

Research

A seed industry can be launched on the basis of knowledge and experience gained in other countries, but problems inevitably arise which cannot be solved in this way. Scientists and laboratories have to be provided to solve these problems as they arise, and to foresee and forestall difficulties before they become acute. The need for research in plant genetics has already been mentioned, but a research service to cover the whole field of seed technology is necessary. Examples of subjects that may require investigation are flowering and seed setting, isolation, seed dormancy, special testing techniques, seed-borne diseases, mechanical damage to seed, storage and senescence. Many of these problems fall within the field of expertise of the specialists working in seed-testing stations, and these establishments are in a good position to recognize and formulate problems and to obtain material for research. The best location for research activity is therefore in association with a large seed-testing station.

Quarantine

A minor (but essential) requirement is a quarantine service to watch over imports of seed from other countries and guard against the introduction of exotic seed-borne diseases. A small quantity of seed imported, for example, by a plant-breeding station can be tested by germinating all the seeds in an isolated glasshouse and observing symptoms in the growing plants. This procedure is obviously impossible for large consignments, which can only be assessed by the testing of samples in a laboratory. If infected seeds are few in number, they may not appear in the sample, and consequently an important disease may not be detected. Laboratory testing is therefore not an effective procedure, and it is better to restrict imports to seed lots produced in areas known to be free from diseases that have to be excluded. There is no need to prohibit the import of seeds infected with diseases which are already prevalent in the country, but if there are statutory health standards for seed sold to farmers, importation of seed lots not conforming to these standards cannot be permitted.

Extension work

The national extension or advisory service has an important part to play in obtaining the best possible utilization of high-quality seed.

Particularly in the early stages of development, the farmer needs to be persuaded of the superior value of the new cultivars and assured of the high purity and germination standards of certified seed. He must also be made to understand that seed deteriorates on his farm, through contamination or otherwise, and that it is necessary to buy new seed from time to time. New cultivars may not give of their best under traditional farming methods, and sales promotion is not enough. Propaganda should be directed to raising the standard of farming by better tillage, sowing and harvesting methods, and by increased use of fertilizers and pesticides.

An extensive educational campaign has therefore to be planned and executed by the extension service, using all the available techniques of propaganda. A method that has proved particularly suitable for promoting high-quality seed is the issue to selected farmers of "mini-kits" or "production kits", each kit containing a quantity of seed of a high yielding cultivar, together with a sufficient amount of fertilizer and pesticide, and any other input necessary to maximize the yield.

The campaign should be planned in consultation with seed technologists and agronomists, and its operation synchronized with the seed production programme. Nothing can be worse than failure to meet demand when it arises, or the distribution of seed which does not in fact meet the standards claimed for it. Supply should keep ahead of demand, and there is much truth in the motto of a successful seed company in Africa—*Availability creates demand.*

Philosophy

The development of a seed industry where none existed before should not be regarded as a benevolent operation for the assistance of impoverished farmers. Rather should it be seen as a particular part of a general development plan for the benefit of every sector of the community. The total amount of capital finance necessary in the beginning cannot be obtained without government sponsorship; interest rates may be low, but some return is expected. A well-founded and efficiently-managed scheme does not require annual subsidies for operational purposes. These are paid ultimately by other sectors of the community and should be avoided as far as possible.

Actually, the capital required for fixed assets such as buildings, equipment and machinery is less than for other industries. The main requirement is for working capital. Seed has to be purchased at harvest time, but is not sold until the following sowing season and even then, sales may be on credit with no prospect of cash until next harvest time.

A seed industry, however, does require a high level of technical and commercial management. Seed production is not a routine process, and is far removed from the assembly-line concept of modern industry. The situation changes from year to year—the weather is unpredictable, the pattern of pests and diseases changes, new cultivars are introduced. The product is a living package always at risk, and surpluses are a dead loss. Managers need to be skilled and adaptable, able to cope with the critical situations and emergencies which are typical of the industry.

Production costs for seed are higher than for grain, mainly because of the additional precautions that have to be taken to prevent contamination and loss of germination capacity. In some countries, seed produced under a planned project has been sold below cost price, or given to farmers in exchange for an equivalent quantity of their own seed. Where seed is sold below the price of grain, there is an obvious incentive for farmers to eat or sell their own produce and buy seed, and an artificially high demand is thus created for high-quality seed. This seed, however, may be sown under conditions of traditional husbandry for which "high technology" cultivars are not suitable, and good seed can be wasted in this way. Such cultivars should not be released until the research organization has developed husbandry methods which will allow the cultivars to develop their full potential, and which are acceptable to the more progressive farmers. Release of the new cultivars is thus co-ordinated with the advocacy of new cultural methods by the extension service. Ideally, if a high price is charged for seed, only farmers who are prepared to husband it well will buy it.

Experience suggests that farmers should be charged a price which covers at least the operational costs of production. If farmers are unwilling to pay this price, there is something wrong—either the cultivar is not good enough and the farmer cannot expect a fair return on his outlay, or seed production methods are inefficient. If produce prices are low because of over-production, there is a fundamental weakness in the economic planning. The cost of plant breeding can be recovered through plant variety rights or by the sale of basic seed.

In general, subsidies from taxation should be directed to long-term objectives such as basic research, technical training, season-to-season storage, and the building up and maintenance of reserve seed stocks.

As an exception, it may be in the public interest to subsidize the sale price of seed where a substantial export trade is involved, e.g. cotton, jute, tobacco and groundnuts. Trade in such products is facilitated if the material sold by farmers is of a uniform type, and this can be improved by "flushing out" seed of inferior cultivars propagated year after year by the farmers themselves.

While the development of a seed industry has to be co-ordinated with other sectors of the agricultural and general economic development programme, experience has shown that it is not possible to implement the whole of a seed improvement scheme simultaneously. Improved cultivars have to be produced, their value demonstrated, and leading farmers persuaded to adopt them; these are the first steps towards creating a demand for better seeds. This suggests that the initial phase of the development plan should provide breeding, demonstration and production on experimental farms of limited quantities of seed.

FURTHER READING

Cereal Seed Technology, chapters 6 and 10. FAO Agricultural Development Paper No. 98.

Government of India Seed Review Team Report 1968, Manager of Publications, Delhi.

Law, A. G. & others (1971), *Seed Marketing*, National Seeds Corporation. New Delhi.

Lawrence, W. S. C. (1971), *Plant Breeding*, Institute of Biology's Studies in Biology No. 12, London.

Outlook on Agriculture—Special Seed Industry Issue, Vol. 8, No. 5, 1975 (ICI).

"Project Seed Laboratory 5000", in special equipment number of *Proceedings of the International Seed Testing Association*, Vol. 34, No. 1, 1969.

3

DEVELOPMENT, RIPENING, DORMANCY, AND GERMINATION

Flowering and seed setting

Seeds develop from flowers, but not every flower bud gives rise to ripe seeds, and in forage grasses less than one third may do so. The casualties occur at all stages—buds fail to develop into flowers and are shed, flowers are not pollinated, and fertilized ovules fail to develop into seeds.

Failure of pollination may be due to the lack of suitable pollen or to the absence of the right pollinator. Rain and high humidity are inimical, and more flowers are pollinated when the weather is dry and sunny. Even when overhead conditions are favourable, the lower flowers in a dense crop may not be pollinated; this applies particularly to grass crops which have been lodged early in the season.

The development of buds into flowers and of fertilized ovules into seeds depends on the supply of water, mineral nutrients and light, and for these there is always competition between plants and between different parts of the same plant. If water and minerals are limiting, they will tend to be monopolized by the stronger-growing plants in the crop, while the weaker plants are checked and shed their flower buds or developing seeds. Within a plant, the buds and seeds nearest to the supply, i.e. close to the main stem, continue to develop, while those on the lateral branches abort.

For their development, seeds utilize substances (mainly carbohydrates) synthesized in the leaves. For this material there is competition within the plant, and ovules situated close to the leaves obtain more of it than those that are further away. That is why the seeds in the lower part of a cereal spikelet or ear are bigger than those in the upper part. In a thinly sown crop, light penetrates between the plants, green leaves function at various levels along the stem, and provide carbohydrates which enable adjacent ovules to develop. In a dense crop, however, little light penetrates and photosynthesis is restricted to the terminal leaves. In these conditions, flowers and seeds are developed only on the upper parts of the plants, close to the source of carbohydrates.

Photosynthesis can take place in any green part of the plant and, in fact, a substantial quantity of the food material that is used to build up the seed comes from adjacent non-leafy organs. Examples are the awns of barley, the glumes of wheat, the husks of maize, and the fruits of pulses and brassicas. This is possible only if they are situated on the upper parts of the plant where there is sufficient light. At an early stage of development, the seed may be unable to utilize all the carbohydrate available, and in the case of pulses the surplus is stored in the enveloping pod and transferred to the seeds later.

Development and ripening

A seed develops from an ovule situated inside the ovary of a flower. The ovary contains one ovule, as in cereals, or several as in pulses. While the ovule develops into a seed, the ovary develops into a fruit. Typically, three stages can be distinguished in the development of the seed after pollination.

1. *Development of the embryo.* After the sexual fusion, there is rapid cell division and by the end of this stage the embryo is almost fully formed. The moisture content throughout is about 80 percent.

2. *Accumulation of food reserves.* These are manufactured in the green parts of the plant and are transported to the developing seed. Table 3.1 shows how the maize caryopsis continues to increase in weight after the embryo has developed sufficiently for complete germination.

Seeds can be divided into two types according to the internal location of the food reserves—*endospermic* and *non-endospermic* seeds. In endospermic seeds, the reserves are deposited outside the embryo to form the endosperm. During this stage of development, the embryo itself makes little growth, but the seed as a whole gains weight because of this added material. When the seed subsequently germinates, the embryo absorbs the food material necessary for rapid growth from this endosperm. The seeds of cereals and other members

Table 3.1 Growth of maize caryopses after pollination (from Walker, *J. Sci. Agr.* **13**, 642. (1933))

Days from silking	13	21	31	43	55
Caryopsis weight (mg)	34	59	135	190	201
Germination (%)	22	72	99	98	98

of the grass family are endospermic. In non-endospermic seeds, the incoming material is immediately absorbed into the embryo and stored in special leaves called *cotyledons*. When the seed is fully formed, the entire space within the seedcoat is taken up by the embryo and there is no endosperm. Examples of non-endospermic seeds are pulses and brassicas. In some species, e.g. cotton, the absorption is not complete and there is a rudimentary endosperm, but this is exceptional.

During this second stage of development, the dry weight of the seed increases three times or more, and the moisture content falls to about 50 percent. Any increase in the size of the embryo is due to enlargement of the cells formed in the first stage, rather than to further cell division. At the end of this stage, the seed is structurally complete.

3. *Ripening*. During this phase the seed dries out. There is little or no increase in the material content and the dry weight remains constant, but the moisture content falls to somewhere between 10 and 20 percent. Finally, a layer of cork is laid down at the base of the seed. This severs the connection with the mother plant, cutting off the water supply and forming a point of structural weakness at which the ripe seed can easily be detached. The time required for this stage is very dependent on weather conditions.

Loss of moisture is accompanied by colour changes in the seed and fruit, the chlorophyll being lost and the colour turning to something in the yellow-brown-black range, according to the species. In the cereals and other members of the grass family there are progressive changes in the texture of the endosperm. At first the endosperm is soft and exudes a milky fluid when squeezed (milk ripe); later it becomes tough and waxy (yellow ripe) and finally hard and firm (dead ripe). In maize (and to a lesser degree in sorghum) dark colour develops at the base of the caryopsis during ripening. This is the so-called *black layer* and it can be taken as an indication that the ear is ready for harvesting. Parallel changes occur in the leaves and stems. In wheat and barley, for example, at the milk ripe stage the lower leaves are dead, but the upper leaves are still green; as ripening proceeds, yellowing progresses towards the apex.

The three-stage development programme described above illustrates in a general way how a seed is formed, but there is much variation in detail. The time required for each stage varies from about one to three weeks or more according to species. The ripening phase is very much under the influence of the weather, but the following typical times for the first two stages together have been recorded:

oats	14 days	sorghum, sunflower	35 days
barley	26 days	soya	38 days
wheat	27 days	cotton	65 days

Nature of the seed

What the farmer plants or sows is not necessarily the true seed in the botanical sense. In the case of pulses, brassicas and clovers, for example, it is the true seed that is sown but, in other crop species, the true seed may be enclosed within other structures.

In the grass family, the ovary contains a single ovule and becomes a one-seeded fruit. During development, the fruit wall becomes fused to the testa and cannot be separated from it. This special kind of one-seeded fruit is known as a *caryopsis.* What the farmer plants is either the caryopsis alone, or the caryopsis with various appendages attached. In cereals this sowing unit is often referred to as a "grain" (e.g. wheat, barley, rice, sorghum) or, in the case of maize, as a "kernel". It is assumed that the reader understands the nature of the inflorescence and spikelet in this family.

wheat, rye, maize, sorghum	The planting material is the caryopsis.
rice, barley, oats	The planting material is the caryopsis enclosed between the fibrous lemma and palea.

In forage grasses, the planting material may be—

the caryopsis enclosed by its lemma and palea,
or the whole spikelet,
or a fertile spikelet together with a sterile spikelet.

In the Compositae family, which includes sunflower and lettuce, the sowing unit is a one-seeded fruit, but unlike cereals and grasses, the fruit wall is not fused to the seed.

In some crops, of course, the planting material does not include any true seed. The so-called "seed" is a piece of stem with buds, taken from the previous crop, e.g. potato. This book is concerned only with planting material which includes true seeds.

When to harvest?

Harvesting comprises two distinct operations—first, the cutting operation in which the inflorescence together with some stem and leaf is detached from the basal part of the plant and, second, the threshing operation in which the seeds are separated from the cut material.

If seeds are harvested before the ripening phase, i.e. in stage 2, they are undersized and become shrivelled on drying; they are troublesome to detach in threshing, soft and therefore susceptible to threshing damage, difficult to dry, do not store well, and on germination have low vigour. This low vigour is in some cases associated with the permeability of the cell membranes; semi-permeability develops as the seed ripens on the plant, and it is important not to harvest before this critical point is reached.

If harvesting is delayed and the seeds left on the plant after ripening, some may be lost through shedding, lodging, or the depredations of insects and birds. Seeds which remain on the plant may be over-dried and brittle, and so liable to breakage during threshing. Alternatively, they may deteriorate in germination capacity and vigour due to weathering. For example, in pulses the pods may absorb water in a heavy rain storm and retain it for some time.

At what intermediate point during the ripening phase should seed be harvested? If the two operations are distinct, it is possible to cut early, allow the cut material to dry naturally and then thresh later. In combine harvesting, however, the two operations are carried out together, and cannot start until the seed is dry enough to escape injury.

It may seem best to harvest early and complete the drying artificially, but drying is costly, and sufficient capacity may not be available. If seed is left until it is completely ripe, its water supply has already been cut off, it is easily detached from the mother plant and little artificial drying is required, if any. Maize and wheat provide an interesting contrast. In maize the ear is tightly held within the husks; in this closed space humidity is high, and in hot weather fungi and insects can flourish. The ear is therefore separated from the plant early in the ripening stage while the moisture content is still high; the husks are removed and the ear dried naturally in a crib or artificially. In wheat the ears are more exposed, the air around the seeds is too dry for fungi and insects, and the seed can be left to ripen longer in the field. In general, seed tends to be harvested earlier than grain for feeding in order to minimize deterioration and losses.

In modern cultivars of cereals and pulses, seed ripening is uniform over the whole crop. One reason for this is that the plants are annuals with limited growth. In the crowded conditions of a farm crop, there are few stems on each plant, seldom more than three, and on each stem inflorescences develop at the apex (from the terminal bud) or at the nodes (from lateral buds). The stem stops growing and there are no lateral branches which might bear inflorescences later. The consequence is that all the flowers on the plant open within a few days of each other, and this leads to almost simultaneous ripening of the seeds. Another reason for uniform ripening is that there is very little variation in the flowering times of different plants within the cultivar. It is therefore possible to secure in one harvest operation all the seed at a nearly uniform stage of ripeness, and with some experience it is not difficult to decide when a crop is ripe enough to harvest.

Native cultivars may not show this feature of uniform ripening because, for the small farmer who grows food only for his own family, it is not important and there has been no genetic selection for it. He is content to harvest small quantities of grain as it ripens or according to his household needs, and a harvest spread over several weeks is no disadvantage. For the farmer who is growing a bigger area as a cash crop, however, harvest is a less leisurely operation; it has to be completed within a limited time and with as little effort as possible, and uniform ripening is important to him. It is a feature that breeders have engendered in modern cultivars by selecting for a determinate growth habit and for simultaneous flowering of individual plants.

The purpose of growing forage crops is to produce leaves and stems rather than seeds, and this poses problems for the specialist seed grower. One difficulty is the bulk of leaf and stem that has to be handled and threshed. A grain crop flowers freely, a high proportion of the flowers set seed, and about half the ultimate weight of plant material is seed—but in a forage crop it may be only 10 percent or less. Another difficulty is the spread of flowering time. Forage plants may be perennial or annual, and the form of growth is indeterminate. Each plant has many stems; either the inflorescences arise on lateral branches, leaving the terminal bud to continue vegetative growth or, if the terminal bud produces an inflorescence, vegetative growth is continued from the lateral buds. In tropical grasses, moreover, there are wide differences between the flowering times of individual plants.

Flowering, and consequently the ripening of seeds in forage crops, is spread over a long period. At any time there are present in the crop, seeds at all stages of ripeness. A single harvesting operation secures a mixture of ripe and immature seeds, and many ripe seeds may have already fallen on to the ground. Some loss is therefore inevitable, partly from the shed seed and partly from immature seeds which have to be removed in the post-harvest cleaning process. The aim is to harvest when the number of mature seeds attached to the plants is greater than at any other time during the growing season. This can be judged by the colour and consistency of the seeds, and the appearance of the inflorescence and adjacent stems and leaves, but is best decided on the basis of moisture content determined by local experience and experimentation. A useful general rule is that the earliest formed inflorescences produce more and larger seeds than later ones.

One of the earliest forage crops to be grown in Europe was ryegrass. Seed production became a specialized business, and the seed growers naturally aimed to secure the biggest possible yield of easily harvested seed. This set up a process of genetic selection which over a long period produced strains of the grass with more uniform flowering and higher seed yield, but at the expense of more desirable forage characters such as leafiness and persistence. Over the last half-century, many tropical grasses have come to be appreciated as pasture and forage crops, and a demand for their seed has arisen; but seed production has not been practised long enough for seed productivity to be improved by mass selection. European experience has shown that it is possible for plant breeders to do this only to a limited extent without sacrificing desirable forage characters, and improved seed yields have resulted from more efficient harvesting rather than from breeding.

Nevertheless, the possibility in the future of spectacular improvements through plant breeding cannot be ruled out. A few years ago in Australia a mutation appeared in *Phalaris stenoptera*, one of the canary grasses, which confers retention on the inflorescence of the fully ripe seeds. In this species harvesting is troublesome, because the seeds are shed in succession as they ripen, but if this gene is incorporated by plant breeders, it will be possible to harvest without loss by means of a combine harvester.

Dormancy

Growth of the embryo is arrested in the ripe seed, but starts again on germination. The three universal requirements for germination are water, oxygen and a temperature anywhere in the range 5–45°C. A seed that germinates when these three requirements are satisfied is said to be *germinating-ripe*. A newly ripe seed, however, is not germinating ripe, but is in a state of *dormancy* and fails to respond to these environmental conditions. Unripe seed harvested after the completion of the first developmental stage has the ability to germinate (though the seeds are valueless for sowing), but dormancy appears as the seed develops and ripens. The duration of this state of dormancy is extremely variable, it may last only a few days or for several years. At some time, however, an *after-ripening* process occurs, at the end of which the dormancy has disappeared and the seed germinates freely in conditions suitable for normal plant growth.

What happens during after-ripening is not the same in every species. There may be a visible change; there is no example of this among crop species, but in the shrub *Ilex* the embryo has not developed sufficiently and further growth takes place later. More commonly, there are invisible physical or chemical changes in the testa, endosperm or embryo. Dormancy may be due to more than one factor, and after-ripening may therefore involve more than one process.

In pulses and clovers, hard seeds are produced which are incapable of absorbing water; after a time the testa loses its waterproof character and the seed can germinate. In other cases the testa is impermeable to gases, so that oxygen cannot enter and carbon dioxide cannot escape; eventually the testa becomes permeable and the embryo can respire freely. In other cases the testa or fruit wall forms a physical barrier which constrains the embryo; germination cannot take place until the barrier has softened sufficiently for the growing embryo to break through.

In most species, however, dormancy and after-ripening are chemical phenomena and not fully understood. There seems to be a block to the physiological processes involved in germination. The block may be the presence of an inhibitor or the absence of a hormone or other growth-promoting substance; both inhibitors and promoters may be present, with the former predominating. In after-ripening the balance is changed either by the loss of the inhibitor by leaching, evaporation or oxidation, or by the production of a promoting substance. Even though the chemical processes are not fully understood, it is well known that under certain conditions dormancy disappears. The best-known dormancy-breaking agents are:

leaching
temperature—dry heat, moist chilling or night-and-day alternations of temperature
physical abrasion or softening of integuments
nitrates in solution
carbon dioxide
light

All these conditions occur in nature and any kind of seed will react, quickly or slowly, to one or more of them, though not necessarily to all.

There is also the puzzling phenomenon of *secondary dormancy*. When a seed recovers from its initial dormancy through one of the agents listed above, but finds itself in conditions unsuitable for germination it may develop secondarily a different kind of dormancy which can be broken only by another agent.

Seed dormancy is one of nature's methods of preserving the species. Each year, over most of the world, a growing season alternates with a period that is unfavourable for growth because of cold or drought. Ripe seeds of wild plants fall into the soil at the end of the growing season and do not germinate until rain or warm weather provide suitable conditions for germination at the start of the next growing season. However, conditions during the winter or dry season are not consistently unfavourable for germination and, if there were a brief spell of warm or wet weather and all the seeds in the soil were germinating-ripe, they would all germinate and the seedlings would be destroyed in a subsequent period of frost or drought. If this were to happen, there would be no seed left in the soil to germinate in the following growing season, and the species would cease to exist. Dormancy prevents this happening, because the seeds are resistant to germination and do not respond to the appropriate physical factors; by the time of the next growing season they have after-ripened and are capable of germinating. This after-ripening may have been brought about during the dead season by such factors as low temperature and heavy rain, or at the start of the new growing season by nitrate or carbon dioxide produced in the soil.

There is still another danger. In some years the early rains which have induced the seeds to germinate may be followed by a drought in which the young plants perish before they are able to produce seeds. The saving factor in this situation is that all the seeds in the soil have not germinated; there is still a reserve of seeds which may remain dormant for one more year, or perhaps even longer, before germinating.

In any locality there are dormant seeds buried in the soil. When, for example, grassland is ploughed, the seeds which are brought to the surface germinate and plants of species which have not been seen for years suddenly appear. Similarly, if forest is felled or burned, there is a flush of seedlings from seeds that have been lying dormant in the shade. This long-term dormancy is probably secondary. The principal factor that breaks this secondary dormancy may well be light, to which they are exposed when they are brought to the surface, or when the overhead vegetation is removed, but the sudden change in the availability of oxygen may also be involved.

Seed dormancy is a desirable character in cultivated species also, though not to the same degree as in wild plants. If seeds were germinating-ripe at harvest time, a shower of rain might cause them to germinate in the field, and they might sprout on the mother plant even before cutting. It is essential therefore that the seeds be dormant enough to resist germination until they are dried and safely in store. This applies particularly to groundnuts, which may lie for some time before harvest in soil moist enough to induce germination. Even in store, some resistance to germination is desirable, particularly in humid regions where seeds may absorb moisture from the atmosphere. It is significant that in the wet tropics the principal food crop (rice) has a higher degree of dormancy than other cereals.

In farm crops, seed dormancy must not be too persistent. Seed should have after-ripened and become fully germinating-ripe by the time of the next sowing season. There is no need for the kind of dormancy that persists for several years and is found in wild species; indeed it would be a disadvantage.

The plant species that have been chosen and developed by man to be cultivated for their seeds are those which have the right degree of dormancy. This character is genetically controlled and, within a species, the cultivars that have been selected, first by farmers and later by plant breeders, have sufficient dormancy for the conditions in which they are grown. There have been examples of apparently excellent cereal cultivars being hopefully transferred to a region of higher rainfall, but proving quite unacceptable because of sprouting at harvest time.

Dormancy is due, in part at least, to inhibitory substances which develop during ripening in the field, and the amount of inhibitor produced seems to be influenced by weather conditions. In warm dry weather, relatively little inhibitor is produced, and with its disappearance after-ripening is soon completed. In unfavourable weather, however, more inhibitor is produced and dormancy after harvest is prolonged. In groundnut cultivars, the persistence of dormancy during storage can depend very much on temperature; it may persist for months at 5°C, but disappears in two or three weeks at 35°C. In grasses, a high proportion of seed may still be dormant at sowing time, but this dormancy is broken directly after sowing by nitrate in the soil water, by the daily alternations of temperature, and by the exposure of imbibed seeds to light. The only form of dormancy that is troublesome in practice to farmers and seedsmen is hardseededness in herbage legumes. This can be overcome by abrading the outer layers of the testa (a process known as *scarification*), but this involves the risk of damage to the embryo within.

Dormancy in farm seeds is more of a nuisance to the seed analyst, who has to carry out germination tests soon after harvest, while the seeds are still dormant. To induce the seeds to germinate in laboratory conditions he applies one of the dormancy-breaking agents mentioned above—leaching, alternating temperatures, nitrate, moist chilling, and so on. The procedures recommended for each species are given in the International Rules for Seed Testing.

Longevity

In nature the ripe seed falls to the ground, but because of dormancy does not germinate even if imbibed with water. In some cases viability may be lost within a few days, e.g. the seed of poplar trees, but in general the seeds of wild plants and weeds can remain in the soil for long periods without losing viability. In the family Leguminosae, many seeds are hard and do not absorb water, but the seeds of other families become fully imbibed. *Chenopodium album* is a common weed in Europe; its seeds have been reported as having a germination capacity of 70 per cent after being buried in the soil for 24 years, and occasional seeds believed to have been buried for centuries have been found viable.

In contrast, the seeds of cultivated food plants are incapable of remaining viable in the soil for long periods. After-ripening is completed in a few weeks or months, and they germinate in the following growing season. Seed is stored in dry conditions primarily to check the growth of moulds. This, of course, prevents germination after dormancy has disappeared but, at normal temperatures, viability is maintained for only a limited period. There are records of seeds maintaining viability for many years, mainly hard seeds of the family Leguminosae, but these were no more than a few survivors from the lot of seed originally stored. The high germination capacity that is necessary for agricultural seed is rarely maintained for more than 3 or 4 years, and some species cannot be stored in normal conditions for more than a year.

Death of the embryo in dry storage is not due to the exhaustion of food reserves; these are abundantly present in the endosperm or cotyledons, but in the absence of water cannot be utilized. It is rather the end of a deterioration process which can start when after-ripening has been completed and continues at a rate which is dependent on the temperature and moisture content. This deterioration is due to biochemical and physiological changes, and possibly the breakdown of the chromosome material. These changes show themselves in various ways, such as lower respiration rate, loss of enzyme activity, leaching of solutes when the seed is imbibed, partial necrosis, abnormality and decreased vigour in seedling development. It is thought that in an imbibed seed such changes are normal, but there is sufficient physiological activity for the damage to be repaired. In a dry seed this does not happen; the deterioration is progressive, the damage becomes irreversible, and ultimately results in complete loss of viability. Experimental work has shown that if seeds taken from dry storage are partially moistened and kept for a time at a moisture content too low for germination, then when they are eventually sown they may show enhanced vigour as compared with seeds sown direct from storage. It may be that this is due to the opportunity provided to recover to some extent from the deterioration that has already occurred.

Dry seed can be stored for an indefinite period at very low temperatures and hermetically sealed. Refrigerated storage is costly, except for small quantities, and this method is used by plant breeding stations and certification authorities for preserving germ plasm and type specimens of cultivars; it is not applicable to large quantities of farm seed. It has been suggested that if dormancy could be prolonged by chemical means, this might facilitate storage in difficult tropical conditions.

Figure 3.1 illustrates what can happen to a seed from the time when it is borne on a plant until it ultimately germinates to produce a seedling, and the differences between wild and cultivated plants.

Germination

The embryo within the seed is a plant in miniature. In the ripe seed it has stopped growing, but is alive and respiring very slowly. Even in conditions favourable to plant growth, further development is inhibited by the dormancy factor, whatever it may be. Sooner or later, the seed becomes germinating-ripe and, when provided with conditions suitable for plant growth, the embryo will then resume its development; this is the phenomenon known as *germination*.

The growth factors essential for germination are water, oxygen and a suitable temperature. Water is needed to raise the moisture content of the embryonic tissue to the level of 80 to 90 percent, which is normal for active tissue in growing plants. To achieve this condition, however, the overall moisture content of the seed, including endosperm, may be less than 50 percent. Oxygen is required for the enhanced respiration that accompanies the reactivation of the embryo. With absorption of water, the embryo and endosperm swell, and the consequent rupture of the outer coverings allows free access of oxygen for respiration. The temperature requirement for the initiation of embryo growth is less than for vegetative plant growth, which generally has an optimum temperature of about 35°C. Most seeds seem to germinate best at a temperature about 20°C, but seedling growth responds to higher temperatures. The need for oxygen is well illustrated by rice, a plant which is adapted throughout its life cycle to secure sufficient oxygen in waterlogged conditions. When the seed germinates under water, the only visible development is elongation of the coleoptile sheath which surrounds the plumule. This elongates towards the water surface like the "'schnorkel"

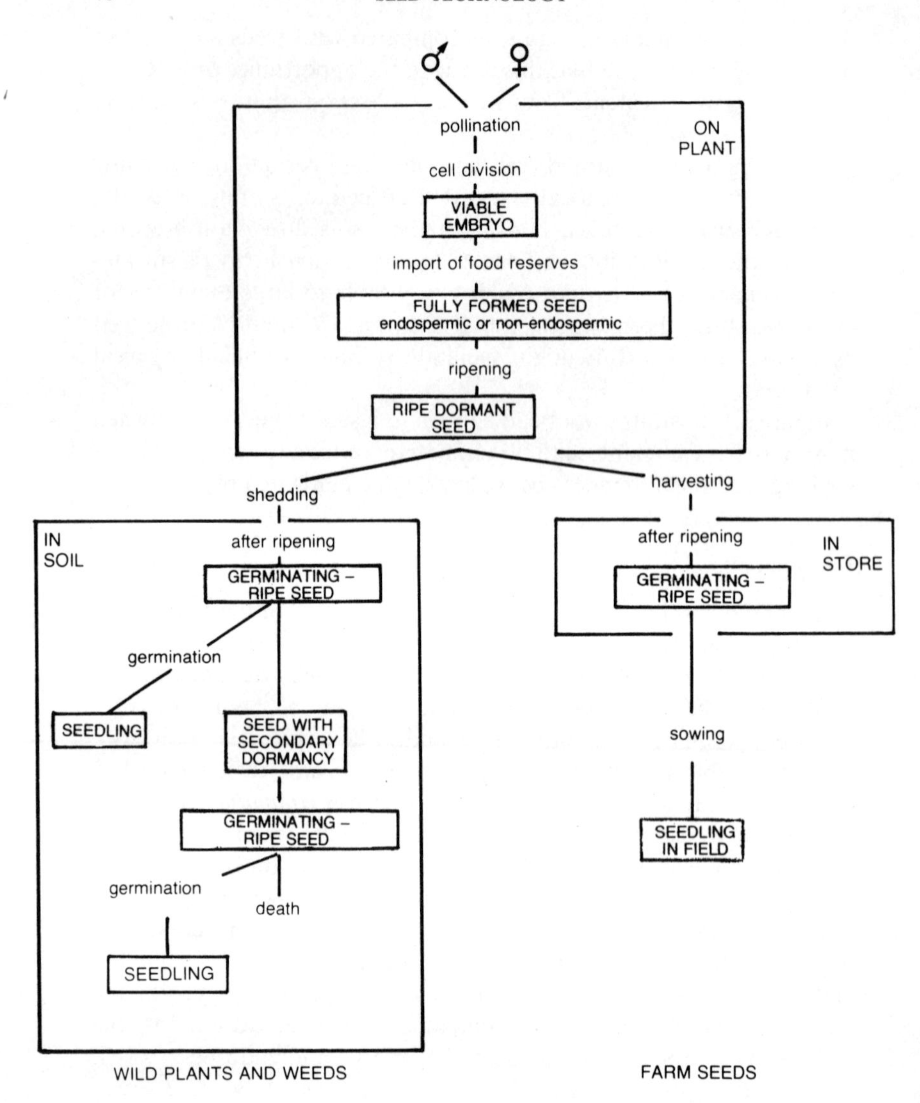

Figure 3.1 Ripening, dormancy and germination in seeds of farm crops, weeds and wild plants.

breathing tube used by a diver, and only when its tip emerges into the atmosphere do the roots and the enclosed plumule begin to elongate. In these conditions, however, germination is slow and development is often abnormal, and seeds are not in practice sown under water.

Successful germination requires the growth of the embryo into a small plant with a root system in the soil capable of absorbing water and dissolved mineral nutrients, and with a green leaf surface capable of photosynthesis. When this stage of development is reached, the seedling is established as an independent plant and, given a favourable environment, is capable of growing into a mature plant.

The embryo is provided with a store of food material on which it can draw until it has grown sufficiently to become independent. This material consists of substances with large insoluble molecules, such as starch, protein and oil. To be used in growth, these substances have to be broken down by enzyme action into small soluble molecules such as sugars, amino acids and fatty acids, and carried in solution to the growing points.

The most significant part of the embryo is the *cotyledon*. There may be one or two cotyledons, and this difference is associated with very striking differences in the adult plants, so much so that the section of the Plant Kingdom which produces seeds inside closed fruits is divided on this basis into two classes—Monocotyledons and Dicotyledons. In the monocotyledons the seed is endospermic, and in cereals and grasses the cotyledon is called the *scutellum*. In the dicotyledons the seed is endospermic in some species and non-endospermic in others.

In endospermic seeds the embryo is small, and most of the space inside the seed is filled with endosperm. The function of the scutellum or the cotyledons is to absorb soluble substances from the endosperm and transfer them to the activated embryo. In non-endospermic seeds the embryo is large, due to the quantity of materials stored in the cotyledons. After conversion to soluble substances, this material is transferred to the parts of the embryo that are actively growing. The embryo has two growing points—the plumule, or terminal bud, which grows upwards to produce a shoot with green leaves, and the radicle which grows downwards to produce a root system.

Germination may be *hypogeal* or *epigeal* (Figure 3.2). In hypogeal germination, the cotyledons remain below the soil and the plumule is carried above ground by elongation of the *epicotyl*, i.e. the stem above the cotyledons or scutellum. In epigeal germination the cotyledons are carried above ground by elongation of the *hypocotyl*, i.e. the stem below the cotyledons. The cotyledons then turn green and function for a short time as foliage leaves, contributing to the growth of the seedling by photosynthesis.

In the cereals and grasses, germination is hypogeal and the scutellum remains below ground in contact with the endosperm, which is contained in the remains of the seedcoat (Figure 3.3). Castor is a dicotyledonous plant with an endospermic seed, but germination

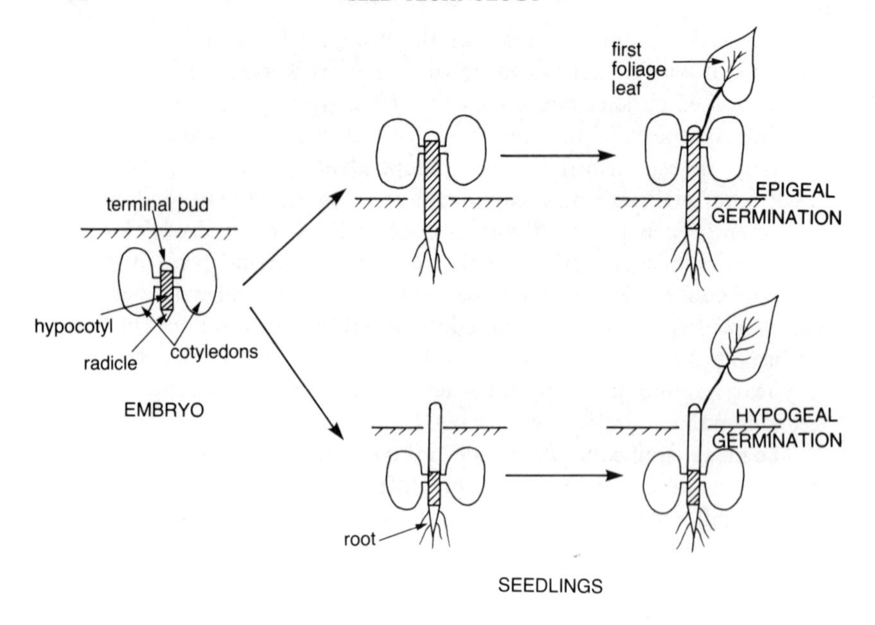

Figure 3.2 Germination of non-endospermic seeds, epigeal or hypogeal according to species (dicotyledons).

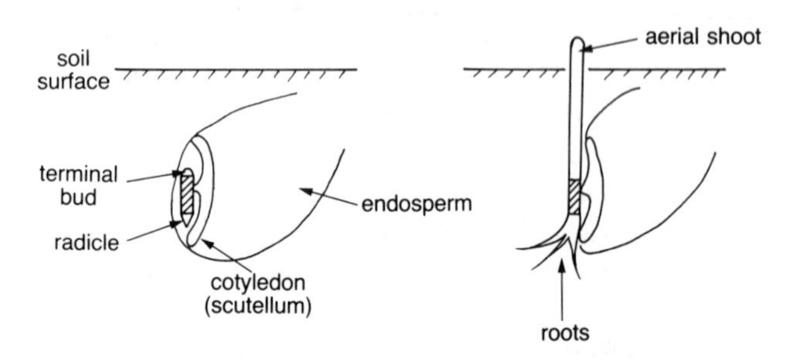

Figure 3.3 Germination of a cereal seed, endospermic and hypogeal (monocotyledon).

is epigeal and, after the endosperm has been absorbed, the two cotyledons are withdrawn from the empty seedcoat and carried above ground to function as green leaves (Figure 3.4).

If the seed is non-endospermic and the germination is hypogeal (Figure 3.2), the cotyledons gradually shrivel underground as the reserve material which they contain is transferred to the growing points. Peas and beans of the genus *Vicia* are examples of this kind of germination.

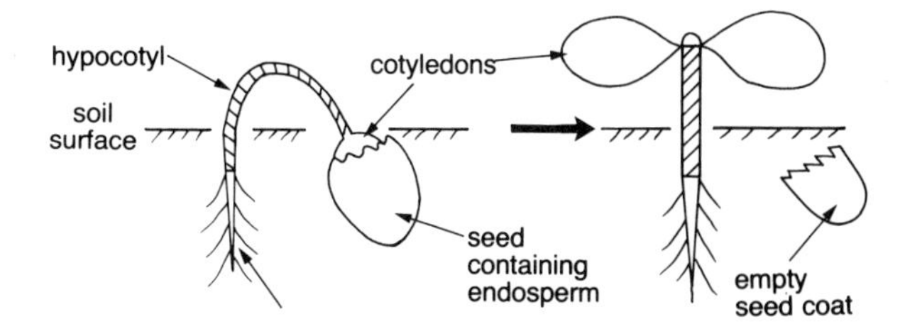

Figure 3.4 Germination of endospermic seed (dicotyledon).

If the seed is non-endospermic and the germination is epigeal, the cotyledons are immediately carried above ground, where they turn green and perform the double function of supplying reserve material to the developing seedling and producing more carbohydrate by photosynthesis (Figure 3.2). Development of normal foliage leaves follows soon after. Examples of this kind of germination are groundnut, soya, cotton, clover and lucerne.

FURTHER READING

Barton, L. V. (1961), *Seed Preservation and Longevity*, Leonard Hill, London.

Boonman, L. J. (1971), "Experimental studies on seed production of tropical grasses in Kenya," *Netherlands Journal Agricultural Science*, Vol. 19, pp. 23 & 237.

Griffiths, D. J. and others (1973). "The seed yield potential of grasses," Welsh Plant Breeding Station Report for 1973, p. 117.

Roberts, E. H. (1972), *Viability of Seeds*, Chapman & Hall, London.

Villiers, T. A. (1975), *Dormancy and the Survival of Plants*, Edward Arnold, London.

4

CLIMATIC AND OTHER REQUIREMENTS FOR SEED MULTIPLICATION

THE DEVELOPMENT OF A PLANT CAN BE REGARDED AS HAVING TWO phases—vegetative and reproductive. In the vegetative phase the plant produces stems and branches, leaves and roots, increasing their numbers and enlarging those that already exist. In the reproductive phase, the plant produces inflorescences and flowers, and pollination takes place leading to the formation of fruits and seeds. These two phases are to some extent antagonistic; when a plant switches from the vegetative to the reproductive phase, vegetative growth becomes less luxuriant and may even stop altogether. An annual plant starts in the vegetative phase, switches to the reproductive phase and then, having produced ripe seeds, dies. A perennial plant alternates between the two phases according to the season. The phase of growth at any time is determined by environmental factors, the most influential of which are light, temperature and water supply.

Some crops are grown for their fruits or seeds, e.g. cereals and pulses. Other crops are grown for their vegetative parts, e.g. beet for roots, grass for leaves, and jute for stems.

In the case of cereals and pulses, the crop is grown by farmers for the production of seeds for consumption, therefore normal agricultural conditions are suitable for the multiplication of seed for sowing. So, seed multiplication and grain production can be carried on in the same region, thus avoiding the cost of transport of large quantities of seed over long distances. Seed multiplication should be in the most favourable districts, so as to obtain the highest possible multiplication rate, full expression of cultivar characters and good harvest conditions. There may, however, be difficulty in securing adequate isolation.

In the case of crops grown for their vegetative parts, the conditions which promote good growth of stems, leaves and roots are not conducive to seed production, the most important factor being rainfall. Seed production is possible in the cropping areas, but there is a tendency for commercial seed production to be carried out in regions of low rainfall. In the United States, herbage seed production is mainly in the drier states on the west side of the country. In Europe, herbage is a more important crop in the north than in the south, but there is proportionately more seed produced in the warmer and drier Mediterranean countries, and some seed is even produced in sub-tropical Africa. For herbage seed, the quantities required are relatively small, and transport costs are not significant. When seed is produced outside the region where it is to be sown, there is a risk of genetic shift (discussed in chapter 10) and this has to be guarded against.

In deciding the location of seed production enterprises a number of factors have to be taken into consideration—climatic, agronomic, biological, social and economic. The climatic factors are light, temperature, rain and wind. It is unusual for a country to be completely uniform climatically and the most favourable region has to be selected. If suitable conditions cannot be found, multiplication in another country may have to be considered. This opens up the possibility of international co-operation.

Light

In many plant species the initiation of flowering is influenced by day-length. Indeed, the switch from the vegetative to the reproductive phase can occur only at the season when the days are of a particular length and, if the plants are kept in the wrong day-length, they remain vegetative indefinitely. This is well illustrated by the results of a simple experiment in which plants of *Dolichos lablab* were grown in pots with different daily exposures to light. The natural day-length was curtailed by covering the plants before sunset and lengthened by electric light after sunset. The results were:

Plants kept in 11 hours light per day flowered in 56 days.
Plants kept in 12 hours light per day flowered in 83 days.
Plants kept in 13 hours light per day did not flower in 100 days.

Crop species of temperate regions tend to flower in the long days of summer, while tropical crop species usually require a shorter day. Generally, the day-length requirement is typical of the species as a whole but, within agricultural species, cultivars adapted to different day-lengths have developed. Wheat includes cultivars requiring either long or short days, which explains why its cultivation is so widespread, and the breeding of cultivars adapted to longer days has enabled the growing of soya in America to be extended northwards. The dates of the start and finish of the tropical rainy season depend on latitude. In Nigeria it has been found that, at any particular latitude, the local cultivar of sorghum has a day-length requirement which causes the plant to flower at such a date that the seeds become ripe after the rains have ceased. Dry weather for harvesting is thus assured. This situation had developed through the selection by farmers of seed from ears that have ripened after the rains.

Jute plants in the vegetative phase are tall and unbranched, and ideal for fibre production. In the reproductive phase, however, they are short and branched, and unsuitable for fibre. For fibre production, therefore, it is important to have the crop in the vegetative phase, and studies in Bangladesh have shown that this requires a sowing date in April or May when the day-length exceeds ten hours. Earlier sowing, before the day has reached this critical length, gives rise to plants which are reproductive and useless for fibre. For seed production, sowing is later, and the plants switch to the reproductive phase when the day-length falls below ten hours in September.

The day-length requirement is not necessarily the same for fruit development as for flowering. In soya and *Phaseolus*, fruit development may require a shorter day, and it has been observed that certain cultivars, when grown under conditions of shortening day-length, may flower for weeks before any fruits develop.

From all this it is clear that the area chosen for seed multiplication must have, at the right time of year, a day-length that is appropriate to the cultivar being multiplied. This may preclude the possibility of producing seed in a location differing in latitude from the region where the seed is to be sown for crop production.

Another variant of light in relation to crop growth is the intensity of radiation, which is dependent on the amount of cloud cover throughout the year. The effect on vegetative growth is more marked in short-day than in long-day conditions, and tropical species respond better than temperate species to high radiation. Other effects, however, are universal; sunshine provides suitable conditions for pollination, and promotes the drying and ripening of seeds.

Sunshine is determined very much by local geography. In Europe, the British Isles have much more cloud cover than nearby continental areas. Studies in Malaysia, which extends over only six degrees of latitude, have shown in detail how much local variation there can be. The number of hours of bright sunshine per day increases by 20 minutes for every degree of latitude north of the equator, and in the lowlands decreases by 10 minutes for every kilometre distance from the sea. There is also a decrease in sunshine with increasing altitude.

Temperature

The switch from the vegetative to the reproductive phase may also be influenced by temperature, and some crop species and cultivars have a critical temperature requirement. Flowering might not follow immediately after exposure to this temperature, but some time later.

Certain cultivars of wheat have a chilling requirement and are sown in early winter; their low-temperature requirement is satisfied at an early stage of growth, but ears do not appear until some months later. If these cultivars are sown after the cold weather has passed, the plants remain vegetative and no ears are formed. Other cultivars are available, however, which do not have a chilling requirement and can be sown in the warm days of spring.

Temperature affects crop development in other ways which can influence the yield and quality of seed. Temperature at sowing time, particularly in the soil, influences establishment. For example, field establishment of *Phaseolus* beans is very slow below 10°C and maize is particularly sensitive, requiring a temperature of at least 12°C. Growth during the vegetative phase is influenced by temperature and determines the final yield. Flowering, pollination, seed setting and ripening are all favoured by warm weather at the appropriate time, and ripeness at harvest is more important for seed crops than for food crops, e.g. the optimum conditions for the ripening of wheat are a temperature of 18–19°C for 6–8 weeks. Too high a temperature, however, may inhibit the development of ovules and fruits, and cause shedding of flower buds or young fruits, as in pulses and cotton. At high altitudes or latitudes, developing seeds may be damaged by early frosts.

Rainfall

Ideally a seed crop requires ample rain during the vegetative phase, followed by a relatively dry period for the reproductive phase. In a dry area it may be possible to supplement rainfall by irrigation; this has the added advantage that it is possible to control the water supply, thus avoiding excesses as well as deficiencies.

Rainfall influences not only the water supply to the roots, but also the atmospheric humidity. Flowering, pollination and seed setting are helped by a moderate humidity, but drier atmospheric conditions are necessary for subsequent ripening. If the humidity is too high, artificial drying of the seed may become necessary.

Temperature and rainfall act together in providing favourable conditions for the reproductive phase, but different crops may have different requirements. Pollination in the small-grained cereals takes place before the flowers open and is not affected by weather. Clovers and grasses, however, are cross-pollinated and are dependent on good weather, clovers more so than grasses because the pollen is transferred by insects. Grass seed survives wet weather during ripening better than clover seed because it is better protected.

Wind

Strong winds during the reproductive phase can cause severe crop losses through lodging, shattering and shedding of seed. The lodging effect of wind can be aggravated by heavy rain, which soaks the ripening ears, thus increasing their tendency to fall over.

Soil

The soil should be the best available as shown by yields of food crops. It should be fertile, neither acid nor alkaline, deep, well drained to avoid water logging, but retentive enough not to dry out. Such soils are usually of alluvial origin. In a cold climate, a heavy soil is slow to warm up at the start of the growing season, and this can delay early growth and subsequent maturity. For forage crops, however, a highly fertile soil encourages excessive vegetative growth, and for seed production a soil of medium fertility is preferred.

The soil should have an adequate mineral status. For high yields this is normally supplemented by applications of fertilizers containing nitrogen, phosphate and potassium. Some soils, however, have a deficiency of minor elements, and these may be of particular importance for the development of normal seeds. For instance, in peas there is a seed defect called "hollow heart" which is due to a deficiency of boron in the soil on which the mother plants grew, and the "split seed" defect in lupins is similarly caused by a deficiency of manganese.

The soil should be free from soil-borne pests and diseases such as eelworm, and for leguminous crops should have the correct strain of *Rhizobium* bacteria for the development of root nodules.

Biological factors

A population of insects, wild or domestic, may be necessary for pollination. The insects most commonly concerned are bees, and they seldom fly more than $1\frac{1}{2}$ kilometres from their hives. In plant protection operations involving the use of insecticides, the effect on pollinating insects should always be kept in mind.

Seed and grain crops are both liable to losses due to plant diseases, insect pests, and the depradations of wild animals and birds. Because of the enhanced value of seed crops, particularly in the early stages of multiplication, areas where the risks of such losses are unduly high should be avoided.

In addition to these general risks to which all crops are subject, there are certain types of diseases and insect pests which are of special significance for seed crops. Seed should as far as possible be free from seed-borne diseases. These are encouraged by wet conditions, and healthy seed is more likely to be harvested from crops in a relatively dry climate. Insects can act as vectors of disease, e.g. soya and lettuce mosaic are seed-borne virus diseases transmitted from plant to plant during the growing season by aphids. In such cases, if there is a district with a naturally low population of the vector, it should be preferred.

Season

In any particular locality, of course, the climate is fixed, but some choice of season may be possible so as to make the best of it. If there is a choice of sowing dates, a time should be chosen which will provide the best possible conditions for the reproductive phase. Any loss in yield so incurred may be compensated by improved quality in the harvested seed.

In Kenya, *Panicum* and *Pennisetum* millets for seed are sown late in the rains to ensure dry weather for harvest.

In the Gezira area of Sudan, there is no choice of sowing time for cotton, but a study of crop development has indicated the best time to harvest for seed. The crop is sown in August, rainfall being supplemented by irrigation. Flowering starts in October, but at that time the crop is under stress due to high temperatures and to limited availability of irrigation water and soil nitrogen, and many flowers are shed. Maximum flowering develops in November, resulting in maximum ripening of bolls in January and February. Bolls formed later are liable to be heavily infected with disease and infested with bollworm due to a build-up during the season. Seed for sowing the following year's crop is therefore taken from the January and February pickings.

The possibility of sowing at various dates should be considered. This practice lengthens the harvest period and spreads the load of work on harvesting machinery and drying equipment. A similar idea is behind the practice in England of growing a number of herbage species with harvest times spread over four months. Tall fescue is ready for harvest in June, red clover in September and other species in July and August. Another advantage of this system is that there are no cross-pollination problems, as there would be if more than one cultivar of the same species were grown on the same farm.

Farms and farmers

For seed multiplication, a district with large farms is best. Where holdings are small and fragmented, isolation is difficult to arrange and small seed lots give rise to administrative difficulties. Isolation, however, can be facilitated by growing seed crops on groups of adjacent holdings. The system of land ownership and tenure should permit continuity, so that crop rotations can be planned well in advance. In tropical countries, the most suitable land units may be estates with perennial crops as their main enterprise; these are managed on a long-term plan, and large isolated areas become available from time to time for annual crops. The farm needs to have the equipment necessary for all operations from sowing to harvest; borrowed implements are liabile to be contaminated with seeds of other species or cultivars. Storage space is needed for the harvested seed, well enough built to provide protection against the weather, and big enough to keep seed lots apart. Farms should be accessible so that extension officers can pay visits, and the harvested seed can be conveniently transported to a processing plant or large-scale store.

Most important of all, however, is the farmer himself. Not only should he have had long experience of food crop growing in his locality, but be especially:

intelligent—to understand the special procedures and precautions necessary in growing crops for seed,
energetic—to ensure that operations are carried out on time,
meticulous—to ensure thorough cleaning of implements and equipment,
reliable—a characteristic which can perhaps be most readily judged by his credit rating at the agricultural bank.

He needs to be supported by technical advice from the extension service, from a cooperative or from a seed company, and such advice should be available to him locally, preferably through frequent visits, particularly in the early years.

FURTHER READING

FAO (1961) *Agriculture and Horticulture Seeds*, chapter 2, Rome.

5

POLLINATION

FOR DEVELOPMENT OF A SEED, FERTILIZATION OF AN OVULE IS NORMALLY necessary, and this follows on the deposition of pollen on the stigma. The pollen may come from:

the same flower, or another flower on the same plant. This is *self-pollination*.
the flower of another plant. This is *cross-pollination*.

In some grasses the seed develops without pollination. This phenomenon is known as *apomixis* and for practical purposes can be regarded as the equivalent of obligatory self-pollination.

Within a species, one mode of pollination tends to predominate, but both self-and cross-pollination can occur. Self-pollination gives a greater certainty of fertilization and is the norm for most of the major food crops of the world. In cross-pollinated species, fertilization is more dependent on external factors such as weather and the activity of insects. In unfavourable conditions seed set can be very low, e.g. in red clover below 50 percent.

In crop species pollen is transferred from one flower to another through the medium of wind or insects. In some species, such as the onion, pollen may be either wind-borne or insect-borne but, in general, each species is adapted to one or the other mode of pollination. Insect-pollinated plants have flowers which by their size, colour, shape or smell, are attractive to insects, e.g. cotton and beans. By contrast, wind-pollinated plants, such as cereals and grasses, have small inconspicuous flowers. Whether adapted to wind or insect pollination, a species may be predominantly either self- or cross-pollinated.

Wind pollination

Pollen can be carried over long distances by wind, and there is evidence of pollen grains being borne from Scandinavia to the Shetland Islands and from Australia to New Zealand. Nevertheless, very little is actually carried more than a few miles, the distance depending on the strength and turbulence of the wind, the topography of the land and the vegetation.

The grass family is an example of a group of plant species particularly well adapted to air-borne pollination. The flowers are enclosed between inconspicuous fibrous glumes and have no coloured petals (figure 5.1). At flowering time the glumes open out briefly, and the anthers and stigmata emerge. The anthers are on long filaments which allow them to swing in the wind and shake out their pollen; the pollen grains are small and easily air-borne. The stigmata are feathery, adapted to catching passing pollen grains. A stigma is not receptive at the time when the pollen is shed from the same flower. The pollen that alights on it can be from another flower on the same plant or from another plant. At any time the majority of pollen grains in the air are likely to be from the surrounding plants and, indeed, apart from apomictic species, grasses are usually cross-pollinated. In some cereals, however, the pollen is shed *before* the flowers open and,

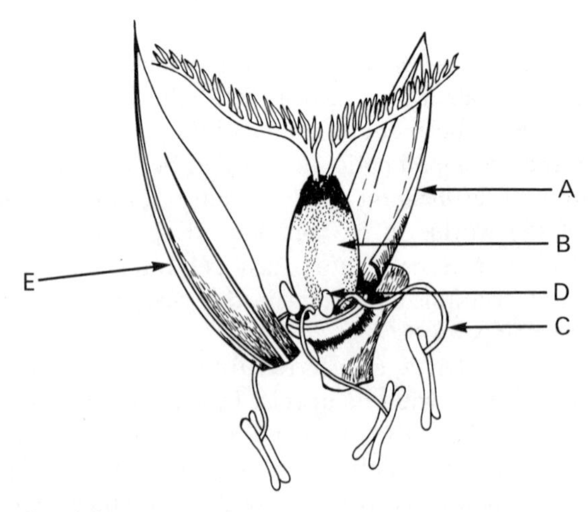

Figure 5.1 Cereal floret.
A. Palea B. Ovary with two feathery stigmata C. Three stamens D. Two lodicules E. Lemma (redrawn from Canada Department of Agriculture).

when the stigmata emerge, they are already dusted with pollen; self-pollination is therefore the rule. In maize, male and female flowers are formed in different inflorescences on the same plant—male flowers in the "tassell" at the apex, and female flowers in the "ears" in the leaf axils. The male flowers open and shed their pollen before the long silky stigmata emerge; cross-pollination is the rule, but self-pollination frequently occurs.

The mode of pollination in the various cereals can be summarized as follows:

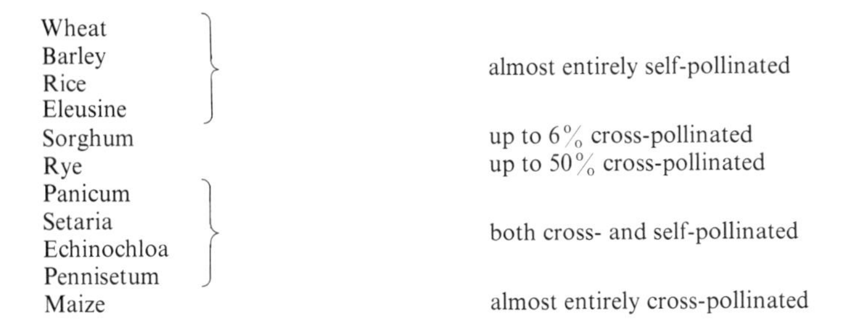

Cereal	Mode of pollination
Wheat, Barley, Rice, Eleusine	almost entirely self-pollinated
Sorghum	up to 6% cross-pollinated
Rye	up to 50% cross-pollinated
Panicum, Setaria, Echinochloa, Pennisetum	both cross- and self-pollinated
Maize	almost entirely cross-pollinated

Pollination for F1 hybrid cultivars

F1 hybrid cultivars have so far been most successful in maize. Maize is normally cross-pollinated but, if self-pollinated by hand, seeds will develop. If this process of self-pollination is continued over several generations, an inbred line is produced which is genetically homozygous. When this stage is reached, it is immaterial whether the pollen comes from the same plant or from another plant of the same line, because every plant in the line is identical. Natural open pollination is therefore permissible in the propagation of inbred lines, provided foreign pollen is strictly excluded.

For the production of seed of a single cross hybrid cultivar, seed of two inbred lines is planted in the same field. The plants of one line have their tassels of male flowers removed, so that their female silks can be fertilized only by pollen from the other line. Seed for sowing is harvested from the detasselled plants only, though seed from the other line need not be wasted and can be used for feed.

Because the pollen-producing line is non-productive, as few plants as possible should be grown, but it is important that enough pollen be produced to ensure complete fertilization. Seed of the two lines is planted in separate rows, usually two rows of pollen plants alternating with four rows of seed-producing plants, but this may be varied according to the capacity for pollen production. The different rows should be clearly marked at the ends to prevent mistakes in detasselling. For successful seed setting, emergence of the tassels and silks has to be simultaneous; this is achieved by planting the pollen line later than the seed line, or by having a relatively late-flowering line for the pollen parent.

Complete removal of the tassels from the seed-bearing line is essential. During inter-row cultivations, side tillers, which would produce late tassels, are removed so that detasselling can be done in one operation. Side tillers on the pollen line, however, are left so as to increase the supply of pollen. Shedding of pollen starts about 48 hours after emergence; the tassel has therefore to be removed one or two days after its appearance, before the flowers have opened, but when it has emerged sufficiently to be easily and completely pulled out of the leaf sheath without damaging the leaves. If done too early, the tassel may be broken, leaving flowers which may open and shed pollen. Damage done to the upper leaves when the tassel is removed can have a significant effect on the development of the ears by reducing the photo-synthetic leaf area. In some lines the tassel emerges clear of the leaves, and this damage is avoided. All the plants do not tassel at the same time, and it is necessary to work over each row, say, 4 to 7 times until the operation is completed. This takes at least 15 days and in this time a man can detassel about 10 hectares. In some countries the workers stand on a mobile platform which is pulled between the rows. Detasselling can be avoided if a male sterile line is used as the seed parent.

Some progress has been made in developing hybrid cultivars of wheat and sorghum, but in these species the flowers are bisexual and male sterility is essential.

Insect pollination

Insects are attracted to flowers in their search for pollen and nectar as food, and the transfer of pollen is incidental to this activity. In some species a highly specialized relationship has evolved between plant and insect, and seed development is impossible without the intervention of one particular insect species. For this mode of transfer the pollen grains are relatively large, their rough surfaces readily adhering to the hairs on the insect's body. The flowers are of such a shape and size that only certain insects can penetrate, and the anthers and stigmata are so situated as to ensure transference of the pollen grains to and from the insect. The main advantage of the insect's involvement in this process is that it provides a means of cross-pollination, but in some species even self-pollination cannot take place without a visit by an insect.

A widespread family with flowers specially adapted for insect pollination is the Leguminosae, which includes pulses among the food crops, and clovers and lucerne among the forage crops. These have butterfly-like flowers and are pollinated by bees, either from hives or wild. Red clover is almost completely self-sterile, and crossing is essential for seed production; nectar is secreted at the base of the long corolla tube and can be reached most easily by the wild bumble bee. Hive bees prefer flowers with more accessible nectar, but will visit red clover flowers if nothing else is available, and seed producers improve seed yields by placing hives in their fields. Lucerne is not completely self-sterile, but no seed can develop until the ovary and anthers are released from the lower part of the flower by a visiting bee. Insects with a proboscis too short to reach to the base of the corolla sometimes gain access to the nectar by biting through the base of the flower, but this does not result in pollination.

In other families the flowers may not be adapted for any particular kind of insect, but are open and readily accessible. Examples are cotton, sunflower, onion, lettuce, carrot and brassicas.

The proportions of self- and cross-pollination and the kinds of insects involved can be summarized as follows for some of the more important crop species:

Lucerne Clovers	Almost entirely cross-pollinated. Bees essential.
Broad bean	Generally self-pollinated, but up to 50% crossing. Bees.
Flax	Normally self-pollinated.
Pea Soya	Normally self-pollinated, but occasionally cross-pollinated by bees.
Cowpea	Normally self-pollinated, but occasionally crossed. Different kinds of insects.
Phaseolus	Normally selfed, but may be up to 30% crossing, depending on species.
Pigeon pea	Normally selfed, but may be up to 15% crossing. Bees.
Groundnut	Normally selfed, but some crossing by thrips.
Chickpea Lentils	Completely self-pollinated.
Dolichos bean	Both self- and cross-pollinated.
Cotton	Mainly self-pollinated, but may be up to 50% crossing, depending on species and country. Bees and other insects.
Brassicas Beet Sunflower	Normally cross-pollinated. Bees, miscellaneous insects and wind. Mainly cross-pollinated by insects, but wind pollination possible.

Protection against foreign pollen

If a cultivar is wholly or partly cross-pollinated, it is necessary to prevent crossing with pollen from another cultivar of the same species. At the time of flowering, a seed-producing crop has to be protected against the entry into its airspace of wind-borne and insect-borne pollen from neighbouring crops and volunteer plants and from wild plants of the same species, e.g. red clover and wild sorghum grasses. In some cases, e.g. brassicas, other species may be interfertile and so have to be guarded against.

The best protection against fertilization by foreign pollen is an abundant supply of the cultivar's own pollen at the time the stigmata are receptive. In these circumstances, the chance of an incoming grain of pollen achieving fertilization is negligible. In practice, however, this has to be supplemented by *isolation*.

Isolation in time may be a possibility. If an early-flowering and a late-flowering cultivar are grown side by side, cross-pollination between them will not occur. In the case of grasses, it may be possible to delay flowering by cutting the crop. In India, isolation in time has been successfully exploited for seed crops of maize; seed crops are grown in the off-season when there are no grain crops in the vicinity.

The kind of isolation that can be most effectively controlled by the agronomist is isolation in space, i.e. separation by distance from sources of contaminating pollen. The actual distance depends on whether the pollen is air- or insect-borne, on the siting of the crop, and on the degree of risk that is judged to be acceptable.

With increasing distance from the source of foreign pollen, the amount of contamination, as shown by the harvested seed, decreases rapidly at first and then more gradually, until beyond a certain distance there is very little advantage in further separation. This is well illustrated by figure 5.2, based on an experiment with ryegrass in the United Kingdom.

The safe distance depends on the direction of the prevailing wind, the presence of trees, high ground or other obstacles to air flow, and the abundance of foreign pollen likely to be released. The area of the seed crop has also to be taken into consideration. In a large crop area, the air and any flying insects are likely to be saturated with the cultivar's own pollen, and insects are more likely to remain within the crop rather than fly from one field to another. As an example, 200 metres may be considered necessary for the isolation of a grass crop of up to two hectares, but for bigger crops, only 100 metres.

The isolation distance required is also influenced by the category of seed being multiplied. High-category seed has a very high purity standard and, because of its value, less risk can be taken than for low-category seed. For example, if isolation of 400 metres is required for the production of inbred maize seed, 200 metres is sufficient for hybrid seed; in brassicas there is a similar difference in the isolation requirements for Basic Seed and Certified Seed. For certified seed, contamination may be more acceptable if the pollen is from a cultivar that differs only slightly, rather than from a cultivar of a widely

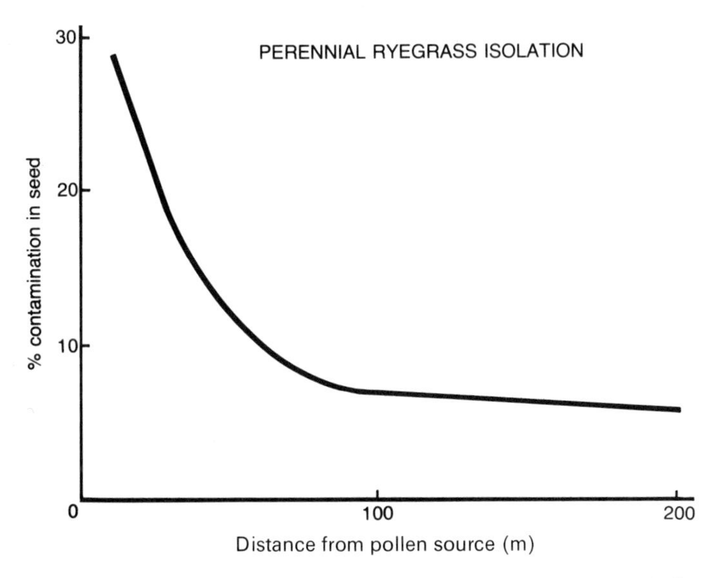

Figure 5.2 Graph showing how the percentage of out-crossed seeds declines with increasing distance from the source of foreign pollen (based on Research and Experimental Record, Northern Ireland 1957).

different type. For example, in the case of upland cotton, a nearby crop is acceptable if it is another cultivar of the upland type, but not if it is Egyptian cotton.

Incoming pollen is most likely to achieve fertilization at the edge of a seed crop, where there is less competition from the cultivar's own pollen. Indeed, wind-borne pollen is liable to be blown away from the windward edge towards the centre of the crop, so that the marginal strip is denuded of its own pollen. The stems and leaves of the crop act as an air filter removing pollen from the air entering the crop. No pollen penetrates into the middle of the crop at this level, but foreign pollen may be deposited by air blowing above crop level. This is illustrated by the results of another experiment on ryegrass in the United Kingdom quoted in Table 5.1. This shows the reduced amount of contamination in the sixth as compared with the first row of the crop nearest to the source of foreign pollen. It is therefore common practice to discard the seed harvested on the edges of a crop, particularly on the windward side. If the edges of a crop are discarded, the isolation distance can be reduced. For example, for the production of certified red fescue seed in the United States, isolation distances can be modified as follows:

No discard	Isolation	46 metres
Discard of 3 metres	Isolation	30 metres
Discard of 5 metres	Isolation	23 metres

Table 5.1 Genetic contamination of ryegrass seed harvested from different rows on the side of the crop nearest to a source of contaminating pollen (from *Proceedings VI International Grassland Congress*, p. 855, 1952)

Distance of edge from source of contaminating pollen	*Percentage of contamination* Row 1	Row 6
8 metres	42	18
30 metres	6	2
120 metres	0.8	0.6

If seed production of a cross-pollinated crop species is concentrated in one district, with the possibility of neighbouring farmers growing different cultivars, zoning arrangements may be necessary to control the siting of crops. These arrangements may be voluntary or legally enforceable. Arrangements of this kind have been made, e.g. for onions, beets and brassicas. A general rule in seed certification schemes is that not more than one cultivar of the same species may be grown on the same farm.

All these factors have to be taken into consideration in deciding what isolation is necessary for any particular crop. In farm conditions, complete protection against foreign pollen is impossible, and some risk of contamination has to be accepted. In brassicas, less than 1 per cent out-pollination has been found in a crop separated from another cultivar by as little as 25 metres, and yet cross-pollination over a distance of 8 kilometres has been reported. The actual isolation requirement is therefore a matter for judgment based on local experience and experimentation but, if too strict, it may be difficult to find suitable locations. A certification authority prescribes isolation requirements and, for each crop species, various distances are set, depending on grade, discard rows and other factors. To encourage seed production, these requirements are kept to the minimum that is considered safe within the region controlled by the authority.

Examples of distances that may be required for some important crop species and the variation that may be permitted are given below:

Crop	Distance
Pea, Cowpea	50 metres for basic seed
Broad Bean	100–200 metres
Lettuce	50–200 metres
Grasses	50–300 metres
Rye, Cotton, Clovers and Lucerne	up to 400 metres
Pigeon Pea, Dolichos Bean, Sorghum	200–400 metres
Maize	up to 600 metres

Brassicas	50–1600 metres
Beet, *Pennisetum*	300–1000 metres
Onion	300–4000 metres
Sunflower	800–1500 metres

No pollen isolation is required for wheat, barley, rice, soya, chickpea, lentil, groundnut and flax.

FURTHER READING

McGregor, S. E. (1976) "Insect pollination of cultivated crop plants," U.S. Department of Agriculture, Handbook No. 496, Washington.

6

AGRONOMY

THE CULTURAL METHODS USED IN GROWING A CROP FOR SEED SHOULD follow the general system of progressive farmers in the same locality, but with the adaptations and special precautions that seed production necessitates. This chapter is concerned with these adaptations and precautions rather than with the details of normal farming practice. Mechanization is common on large farms, whether in state, company or private ownership, but is not essential. The enhanced value of the crop justifies additional expenditure of money and effort, but what is essential for seed production is care and attention to detail, and this can be contributed equally by the small farmer. Vigilance is necessary throughout the growing season, but particularly at critical stages of the crop's development, e.g. at seedling emergence and flowering. The crop can be observed in sufficient detail only by walking through it, and this should not be restricted to the roguing operation. At other times faults should be looked for, their cause diagnosed, remedial action taken if possible and, if not, noted for action next year.

It is important that any application of chemicals for weed or pest control be made at the correct stage of crop development. For this purpose the various stages through which a crop develops have been precisely described for the most important crops (e.g. cereals and soya). Manufacturers of agrochemicals indicate at which of these stages their product can be safely applied without damage to the crop, and their advice should be strictly followed.

Fertilizers

In the vegetative phase of crop development, sufficient mineral nutrients (mainly nitrogen, phosphate and potassium) are required to build up, through the synthetic activities of the leaves, a plant structure with the maximum number of positions at which seeds can develop. After flowering, the active leaf area may be reduced, as in wheat for example, but seed development is dependent on the synthetic processes of the remaining green tissues and for this purpose a continuing supply of nutrients from the soil is necessary.

The application of fertilizers to cereals and pulses should be based on local practice as recommended for food crops, but with any modifications that may be necessary for seed production and appreciating that the value of the crop justifies additional costs. Phosphate and potassium are more important for seed crops than for food crops, especially in the case of pulses, but enough can be applied at sowing time to supply the crop's entire need.

Very heavy application of nitrogen to cereal crops should be avoided as this is liable to encourage foliar diseases, cause excessive vegetative growth and lodging, and may even lead to reduced seed yields, though this warning is more relevant to the long-day conditions of temperate regions than to the tropics. Nitrogen increases the number of tillers competing for the available light, water and soil nutrients, and for material photosynthesized on the main stem. In these conditions many of the tillers die off, while the surviving ones are weakened and do not bear their full potential of seeds.

The lodging which results from excess nitrogen produces a sheltered humid environment unfavourable to pollination. These conditions are made even worse by subsequent rain; many tillers die and are colonized by fungi. Such an environment is unsuitable for the development and ripening of seeds, and the quality of any seed harvested is poor. Another effect of lodging in cereal and grass species is to admit light to the lower parts of the plant and encourage late tillering. These tillers are too late to produce ripe seed, but in their early development compete for photosynthesized material from the main stem and thus reduce the amount available for developing seeds.

Lodging can create harvesting difficulties, resulting in inefficient collection of the seeds that have developed. It should be remembered, however, that in cereal and grass crops late lodging can be beneficial by reducing the shedding of ripe seed when the plants are shaken by wind.

As a general rule the nitrogen should be applied at an early growth stage, but an additional dose may be given later if considered necessary. In leguminous crops, atmospheric nitrogen is fixed by rhizobial bacteria in the roots, but this may not be sufficient for a heavy seed crop, and nitrogenous fertilizers can in some species increase yields significantly. In Malaysia it has been found that a limited application can be justified by increased yield, but anything in excess of this serves only to increase the number of immature and valueless pods produced.

Fertilizer should be taken to the field in clean bags and not in old grain sacks which might introduce foreign seeds. For the same reason, all parts of the distributor must be thoroughly cleaned.

Irrigation

Water can be applied overhead through sprinklers or on the surface through soil channels. Overhead irrigation may encourage foliar and seed-borne diseases and can be inimical to pollination; as a general rule surface irrigation is to be preferred. Surface water should be taken direct from the canal or pump and not allowed to flow through another crop, with the risk of carrying in foreign seeds.

As far as possible, water should be supplied to match four stages of crop development, as follows:

1. Establishment and vegetative growth up to the initiation of flowering—ample water.
2. Flowering—limited water. Slight water deficiency at this stage is believed to promote seed setting.
3. Early phase of seed development—ample water. To ensure the development of the greatest possible number of seeds, it is important that the plant should not be under any form of stress at this stage.
4. Ripening—no water.

Previous cropping

In a seed crop, volunteer plants of a different cultivar or species may appear derived from a previous crop in the same field. These may be plants which have persisted despite drought, cultivation and other hazards, or may have grown from seeds shed on to the ground. Survival of perennial plants can be prevented by thorough cultivation. To dispose of seeds shed on to the ground, the period between crops of the same type must be long enough (allowing for dormancy) for all the seeds to germinate and for all the plants growing from those seeds to have died or been destroyed without shedding another generation of seeds. The actual period required depends on the distribution of rainfall, the incidence of warm weather, and the crop rotation followed. For cereals, an interval of two cropping seasons is desirable, but in some certification schemes one is acceptable. For forage crops, longer intervals are required because of seed dormancy and persistence of the plants. The growing of a seed crop of an annual species on the same land in two successive seasons is permissible, provided the cultivar is the same and the first crop was of certification standard.

A weed species tends to flourish in a certain type of crop and to be checked in others. Wild oat is an annual plant with a life cycle and habit of growth similar to a temperate cereal crop. Its seeds germinate about the same time as the crop seeds, and the resulting plants grow undisturbed and ripen their seeds in harmony. Some of the weed seeds are shed and remain in the soil to germinate in subsequent years, while others are harvested with the cereal grain and may be sown as an impurity in another field. Wild oat plants cannot survive in spaced crops subject to inter-row cultivation or in dense herbage swards which are grazed or periodically cut. Docks are common weeds in forage crops; they are perennial and can survive periodic cutting, but not the harsher treatment of inter-row cultivation. In farming, rotation of crops is an effective way of keeping weeds in check insofar as favourable and unfavourable conditions for any weed alternate. A seed crop should follow a crop which smothers weeds or which permits intensive weeding.

Sometimes plant material is stacked in the field between cutting and threshing. Large numbers of crop and weed seeds are left on the ground under these stacks and the sites should be avoided in sowing seed crops the following year.

Some seed-borne diseases can also persist in the soil, e.g. head smut of maize, clover rot and eelworm. These diseases are kept in check by long intervals between susceptible crops.

Weeds

Weeds are objectionable in all crops because they:

compete for soil water and nutrients.
smother the crop, cutting out light and delaying harvesting.
impede cultivation, e.g. the obstruction of animal-drawn ploughs by rhizomes and strong roots.
impede harvesting, e.g. climbing plants and, if the weed seeds are harvested green, increase drying costs.
may be poisonous.
may be plant parasites, e.g. witchweed and broomrape.
may harbour pests and diseases.

In seed crops, however, they are particularly objectionable because, if they ripen at the same time, their seeds are harvested with the crop and, if difficult to separate, they may be found as a contaminant in the seed, even after processing. Example of such weeds are:

wild oats in cereals
wild red rice in rice
black grass in forage grasses
dodder in clovers
bindweed in kenaf
cranesbill in red clover
suckling clover in white clover

Seeds of this type may be specified in a country's seed laws and designated as noxious weeds.

These weeds are not restricted to crops being grown for seed, but can grow in other crops on the farm. Control measures should therefore be applied to the farm as a whole. The most objectionable is wild oat, the dormant seeds of which persist in the soil. It can be controlled to some extent by the use of herbicides, but for complete eradication roguing is necessary and this should be done on all crops on the farm.

In a seed crop, other crop species should be regarded as weeds, particularly if their seeds are difficult to separate, e.g. one cereal in another cereal, and seeds of gram in a cereal.

The most effective check on the growth of weeds is rapid and uniform germination of the crop seed, giving rise to a dense root system and canopy of leaves. Thick sowing of a crop for seed, however, is not advisable.

The standard methods of weed control in crops are:

1. Drainage, to check weeds which flourish in wet conditions, such as sedge, and to encourage crop growth.
2. Flooding, in the case of rice, to check the growth of weeds which are intolerant of waterlogged conditions.
3. Rotation of crops.
4. Apply farmyard manure only after it has been rotted to destroy weed seeds.
5. Apply fertilizers to promote competitive crop growth.
6. Sow on a clean seed bed obtained by cultivation, pre-watering or pre-emergence herbicides.
7. Sow seed of good germination and free from weed seeds.
8. Destroy weed plants or prevent seeding by inter-row cultivation and roguing. One plant (e.g. *Sonchus arvensis*) can produce 50 000 seeds.
9. Exhaust perennial weeds by repeated cutting and remove rhizomes.
10. Herbicides. These cause less root damage than mechanical cultivation and are preferred in suitable cases.

These methods are all applicable to seed crops, but perhaps the most effective is sowing on a clean seed bed. A seed bed free from annual weeds is obtained by inducing the seeds in the soil to germinate and then destroying the seedlings. However, if the seeds are dormant they will not germinate until later, either in the same year or in subsequent years.

In England, grasses of the genus *Poa* are common weeds in grass seed crops such as ryegrass. The ripe seeds are shed on to the ground and, if after the final harvest they are left exposed to light on the surface, they will germinate and the seedlings can be destroyed; but if they are buried by post-harvest cultivation they become dormant. Effective control can be secured by leaving the soil untouched for about four weeks after harvest. The seedlings of *Poa* and other weeds which have grown and the stubble of the grass crop are then sprayed with paraquat. Subsequently, the dead vegetation is burned and this destroys most of the ungerminated seeds still lying on the ground.

Diseases and pests

The incidence of diseases and pests in a crop is influenced to a certain extent by climate and by their presence in the soil and, as already indicated, this is taken into consideration in selecting districts, farms and particular fields for seed multiplication.

The same basic control methods are used as for food and forage crops. Burial of plant debris by ploughing and rotation of crops reduce the risk of disease being carried over from previous crops. Seed treatment and insecticidal sprays may be normal practice, but special additional insecticides are sometimes necessary for seed crops, e.g. to control seed weevils in clovers and lucerne. The risk to pollinating insects should always be kept in mind.

Some diseases are air-borne or insect-borne, and a seed crop has to be protected against them in the same way as it is protected against foreign pollen, i.e. by isolation. Loose smut disease of cereals is caused by a fungus which produces its spores at flowering time, and these are carried by wind to neighbouring crops. The mosaic and yellows diseases of beet are caused by viruses and are transmitted by aphids, which feed on the leaves; isolation up to a kilometre may be necessary.

Good hygiene in seed stores and granaries can prevent the carryover of pests. Maize kernels are attacked by weevils, which can persist in food or seed stores and emerge to infest the following year's crop. Even if proper hygiene is observed, there is always the possibility of the insects flying from a neighbour's store or crop, and 75 to 100 metres isolation or spraying of seed crops may be necessary.

As a precaution against epidemics of wheat rust, commercial farmers are advised to grow, not one cultivar over a large area, but two or more cultivars with different resistance genes. This method of disease control, however, is not available to the seed grower who, as a precaution against contamination, is restricted to one cultivar.

As in other matters, the high value of a seed crop can justify special measures, e.g. against birds and rodents. The spread of a disease within a crop can be checked by roguing out diseased plants, e.g. groundnut plants infected by the rosette virus.

Seed

The seed to be sown should arrive on the farm in sealed bags ready for sowing, having already received any insecticidal or fungicidal treatment that may be necessary. For the production of seed of F1 hybrid cultivars, seed of the pollen and seed-bearing parents has to be sown in the same field in pre-determined rows, and it helps to prevent confusion if the two kinds of seed are dyed different colours so that they can be easily distinguished. The labels on the bags should be checked and kept as evidence of the cultivar name and seed lot number, in case any defects become apparent. A sketch plan should be made showing the position of the field and the exact area within the field where the seed was sown.

In the case of pulses and leguminous forage crops, it may be necessary to inoculate the seed with the correct strain of *Rhizobium* bacteria, but this need be done only on the first occasion on which the crop is grown in that particular field; thereafter the bacteria will persist in the soil. The inoculant can be mixed with the seed at sowing time.

Sowing date

The major climatic factors influencing crop growth are temperature and water. Crops are sown when these two factors are favourable and can be expected to remain so until harvest time. In temperate regions, moisture is usually sufficient and the sowing date anticipates warm weather; there is only one growing season. In the tropics, temperature is at least adequate for growth throughout the year and, except at times of excessive heat, sowing dates depend on the availability of water, either from rainfall or irrigation. There can be two growing seasons with different day-length characteristics, or growth may be possible throughout the year.

Within a season, little variation in sowing dates for annual crops is possible, and the best local practice for food crops of the same species should be followed. In northern Canada there is a frost-free season of four months or less, and crops cannot be sown until the ground has thawed out. In northern Sudan there is a cool winter season of similar length, and certain pulses cannot be sown until the temperature has fallen sufficiently.

Harvest dates are determined more by the day-length and

temperature requirements of the crop than by sowing date. For temperate cereals this is well illustrated by an experiment in England in 1975 (Table 6.1). Barley was sown at intervals over a period of 83 days, but the difference in date of flowering (which determines date of harvesting) was no more than 18 days. Though delayed sowing had a

Table 6.1 The effect of sowing date on time of flowering in barley. (from East of Scotland College of Agriculture Annual Crop Production Conference 1976)

Sowing date	*Date of flowering*	*Number of days sowing—flowering*
20 February	23 June	124
19 March	26 June	100
22 April	5 July	73
14 May	12 July	59

disproportionately small effect on harvesting date, it did seriously diminish the crop yield. For crops that are normally grown for their vegetative parts (page 33) and for biennial and perennial species (page 73), other factors have to be taken into consideration, and the sowing date for seed crops may be quite different.

Sowing

Preparation of the land for sowing should start in good time to ensure that a suitable tilth is ready by the proposed sowing date. A fine tilth is necessary for small seeds, but not for large seeds such as maize.

The depth of sowing also depends on seed size, large seeds being buried more deeply than small seeds. Depth of sowing may, however, be influenced by the date. If seed is sown in anticipation of the rains, it should be sown more deeply to ensure that only heavy rainfall will penetrate and stimulate germination.

Rice can either be sown direct in the field, or sown in a nursery and the young plants transplanted. Although it is more tolerant of lack of oxygen than other crop species, nevertheless in flooded conditions germination and seedling growth are poor, and it is not until the seedlings are at least 150 mm high that they can withstand submersion. In direct sowing, therefore, the seeds are planted in dry or moist soil, and flooding is delayed. Before sowing in a nursery, the seeds may be soaked and pre-germinated for about three days, and then planted in soil that is moist but not waterlogged. When grown sufficiently, the young plants can be transplanted into a flooded field.

There is always a risk that tractors and implements might introduce seeds from other fields or from other farms, adhering to the wheels perhaps; this should be kept in mind, and precautions taken when necessary. More particularly, there is the risk of foreign seeds being introduced within the seed distributor, drill or planter, and this necessitates thorough cleaning before the machine enters the field.

Except in some forage crops, the seed is usually sown or planted in rows rather than broadcast. The advantages of drilling are that it requires less seed, distributes it more evenly, allows weed control by inter-row cultivation, and provides pathways for roguing and crop inspection. The seed may be sown on ridges—in the case of groundnuts to facilitate harvesting of the buried seed pods or, in the case of crops such as *Phaseolus* and cotton under flood irrigation, to avoid waterlogging, particularly in the early growth stages.

The best distance between rows for any species or cultivar depends on its habit of growth and on the lateral spread of the root system. To allow penetration of light, tall plants are grown in wider rows than short plants, e.g. 750 mm or more for maize, but only 150 mm for wheat. Grasses are relatively short, but side tillers may produce a tufted or spreading habit, and the row width of many species is between 300 and 600 mm. The roots should spread out sufficiently to utilize the soil between the rows, but deeply enough to avoid damage from inter-row cultivation. Because of their meagre root system, the ryegrasses, for example, are grown in rows only 150 mm apart. Within the row, the seeds may be spaced or sown in a continuous line. Maize seeds are planted at least 200 mm apart; small grain cereals are sown in continuous lines, and the machine should have a meter to indicate to the operator the quantity of seed passing from the hopper. This facilitates uniform sowing of the seed over the entire field. If it should happen that the seed runs out, the remaining area should be treated as a separate crop and on no account be sown with unauthenticated seed or with another cultivar.

For the production of F1 hybrid maize, sowing of each line must be completed within five days to ensure uniformity of flowering. Pollen and seed-bearing lines may have to be planted at different times in order to synchronize flowering. To make the pollen rows more easily recognized by workers, a few seeds of another species (such as sunflower) are sometimes mixed with the seed of the pollen-producing line at sowing time, and the foreign plants act as markers.

Seed rate

The quantity of seed sown per hectare determines the density of the plant population within the crop. The minimum rate should be sufficient to provide ground cover when flowering starts, but for weed control ground cover may be necessary at an earlier stage in crop development, and for seed production it may be desirable to modify the growth habit of the plants.

Thin sowing gives widely spaced plants and permits an adequate light intensity within the crop. Numerous lateral branches or tillers compensate for the smaller number of main stems, and together they produce a similar number of flowers and seeds. The soil moisture requirement is less than for a dense crop. A disadvantage of this habit of growth is that the initiation of flowering is delayed, and is later on branches than on main stems, so that the seeds do not ripen uniformly on all parts of the plant; this is particularly noticeable in annual crops and in tropical grasses. The seed rate, therefore, should not be too low. Another disadvantage of thin seeding is that the population of plants may be insufficient to withstand competition by weeds.

Thick sowing gives rise to a dense mass of thin weak stems, with few flowers per stem and poor seed setting. The humid conditions within the crop discourage pollinating insects, promote mould growth, delay ripening and so create harvesting difficulties.

The relation between the amount of seed sown and the yield of harvested seed is elegantly illustrated by the results of an experiment with winter rye carried out in the Netherlands (figure 6.1). Seed was sown at seven different rates ranging from 5 to 180 kilograms per hectare, and the number of plants was in direct proportion to the quantity of seed sown (E). At the lowest seed rate each plant had enough space to spread out and produce about nine stems with ears, but at the heaviest sowing only one ear per plant was produced (A). Nevertheless the number of ears per hectare was greatest at the highest seed rate (F). The number of grains produced in each ear (B), the size of each grain (C) and the yield of grain per ear (D) declined with increasing density of sowing. So, with increasing plant population, the number of ears per plant and yield per ear decreased, but the total number of ears increased. The net result was that the weight of grain produced per hectare increased up to a maximum with a sowing rate of about 50 kg/ha and at heavier seed rates showed a tendency to decline (G).

This experiment demonstrates two points of importance. With increasing plant population the yield of seed per hectare increases up to a point, and beyond that point no significant addition to the yield is obtained. In seed production it is important to secure a high return of harvested seed for every kilogram sown, and the seed rate should be enough to establish a crop which will give a seed yield approaching this maximum. In multiplying breeder's seed, the ratio of harvested

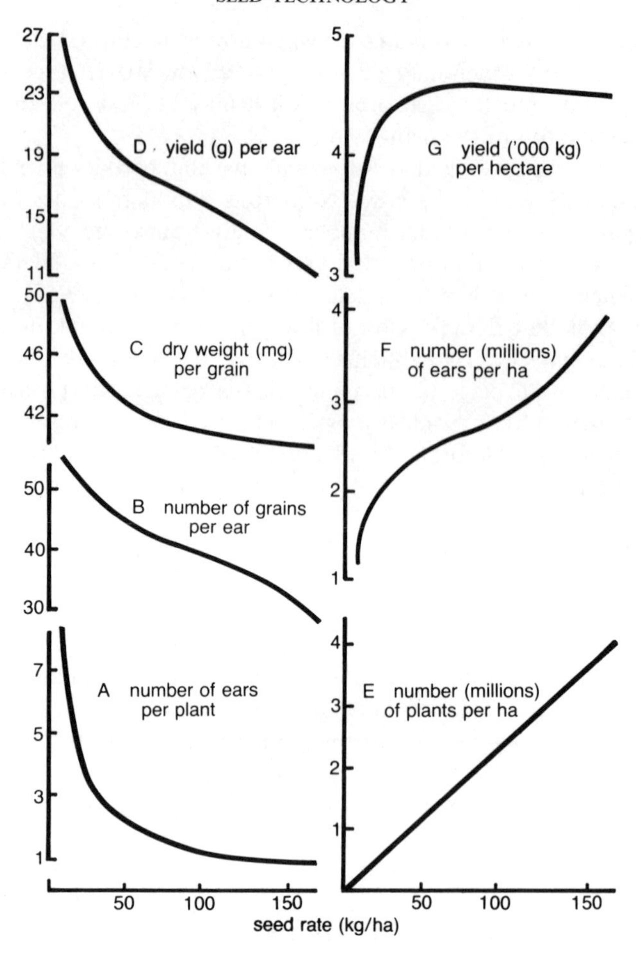

Figure 6.1 Winter Rye—effects of increasing seed rate (adapted from Bruinsma, *Netherlands Journal of Agricultural Science*, Vol. 14, 1966).

seed to sown seed may be more important than yield per hectare, and for this purpose the sowing rate may be quite low. In this particular experiment the multiplication factor ranged from 620 at the lightest seed rate down to 25 at the heaviest sowing. The disadvantages of low seed rates have, however, been noted above.

Another point that arises out of this experiment is that at the higher sowing rates, smaller grains are produced (C). In a sparse population there is more green tissue per ear, and the green colour persists for a longer time (as shown by delayed yellowing of the plants) because of the increased moisture and nitrogen available. More carbohydrates, etc., are therefore synthesized and this enables the grains to develop to a greater weight. As large seed size is a valuable seed quality, this provides one more reason for avoiding excessive rates.

The relationship between seed rate and yield shown in this experiment does not necessarily hold in all circumstances; it can be modified by soil conditions and by crop species or cultivar. African farmers space their sorghum plants widely because there is not sufficient soil moisture to support a denser crop, the actual spacing being based on experience. In these conditions, high density results in a spectacular drop in yield per hectare. In other circumstances, where moisture and nutrients are not limiting, there may be no falling off in yield at the highest seed rates. The relationship depends, too, on the habit of growth of the plants. Species or cultivars with strong branching growth can fill and exploit the space available in a thinly sown crop, and increasing the density produces little increase in yield; this is the situation in crops of *Phaseolus vulgaris*, for example.

In any particular case, therefore, the seed rate should allow for the habit of growth of the species of cultivar, the weed growth expected, and the availability of soil water and nutrients. The aim is to produce a well-spaced population of strong vigorous plants.

Prevention of mechanical contamination

In a previous chapter measures were discussed for the prevention of genetic contamination by foreign pollen. Equally important is the prevention of mechanical contamination, i.e. the admixture of foreign seed. The possibility of this occurring must be kept in mind in all farm operations.

Even if no pollen isolation is necessary, a seed crop should be separated from another crop of a similar type (e.g. another cereal) by a gap of two or three metres, or by a physical barrier such as a wall, a ditch or a hedge. This reduces the risk of plants falling over on to an adjacent crop or of harvesters failing to recognize the crop boundary.

All implements and containers should be thoroughly cleaned between operations. This applies to tractors, cultivating implements, seed drills, cutting machines, combine harvesters, threshers, wagons, lorries, driers, storage bins and so on. After harvest, seed lots must be kept separate in store.

In the case of hybrid maize, unless special harvesting machinery is available, ears from the pollen plants are harvested first and removed from the field.

The most effective means of all, when practicable, is to grow only one cultivar of any species on the same farm.

Roguing

In a seed crop plants may appear which are undesirable because they contaminate the product—their seeds are harvested with the crop seed and, being similar in size and shape, cannot be separated from it or, if they belong to another cultivar of the same species, cannot even be recognized. Such plants, commonly called *rogues*, may not have any significant effect on the productivity of a feed or forage crop, if present in small numbers, but are quite unacceptable in a seed crop. They may be weeds, plants of other crop species, plants of another cultivar of the same species, or *off-types*, i.e. genetic variants which have arisen within the cultivar. Certification schemes set very strict limits to the number of each kind of rogue that is tolerated, depending on the grade.

Roguing is the process of removing these undesirable plants from the crop. Its efficacy depends partly on the distinctness of the rogues, and partly on the skill of the roguers. A rogue can be removed only if it is distinctive enough to be recognized by an experienced roguer walking slowly through the crop. Weeds and other crop species are easily seen, but recognition of another cultivar or an off-type depends on the vividness or the magnitude of the difference. Differences which are so minute as to be seen only with a lens are of no value in field conditions, while others (like flower colour or awn length) are clearly visible. A difference in height or maturity may or may not be apparent; a tall (or early) plant is outstanding in a short (or late) cultivar, but a short plant in a tall cultivar is out of sight. Another factor contributing to distinctness is the uniformity of the cultivar. In a uniform population of plants, such as a pure line, quite a minor difference may be visible. In a non-uniform cross-pollinated cultivar, however, a rogue is not recognizable as such unless it is clearly beyond the normal range of variation. In the extreme case of a synthetic or composite cultivar, roguing may be almost impossible. Roguing is certainly limited in its effectiveness, but it does contribute to the maintenance of purity, and is an essential operation in seed multiplication.

Roguing should be done several times at different stages of crop development. The best time is when the crop is in full flower; at this stage cultivar characters are most fully expressed, and differences are most obvious, e.g. the red rice rogue can be recognized only at flowering. In cross-pollinated crops, however, roguing should be at an earlier growth stage before pollen is released. In maize, rogues are more easily observed when the crop is below shoulder height.

In roguing, particular attention is paid to those parts of the field where rogues are most likely to be found, such as gateways, the sites of old stacks, and places where animals may have been fed. Otherwise, the roguer walks slowly (not more than three kilometres per hour) backwards and forwards through the crop. He has a bag slung from his neck in front of him and scrutinizes the plants in a strip about two metres wide. Each rogue as it is observed is uprooted, so that no part is left to continue growth, and it is placed in the bag. This material is removed from the field and destroyed by burning. To maintain a straight line and minimize damage to the crop, the roguer walks between the rows. It is a help if rows are left blank at intervals of four metres; these provide pathways and prevent overlaps and omissions.

Biennial root crops

Some crop species which are grown for their vegetative parts are biennials, flowering only in the second growing season. Beet is grown for its roots, and within the same species *(Beta vulgaris)* there are types suitable for human consumption, cattle feeding and sugar production. The roots of carrot, parsnip, swede and turnip can be used as vegetables or as cattle food. In cabbage and onion, the leaves grow compactly together to form heads or bulbs, and these are used as vegetables.

For vegetable production seed is sown early in the growing season and within 2–3 months the plants have developed short stems or swollen roots which can be harvested, but no flowers. The switch from the vegetative to the flowering phase requires a period of low temperature and, if the plants are left in the ground over winter, the stems elongate to produce inflorescences in the following season. Seed production, therefore, involves three stages:

1. A warm period favouring vegetative growth.
2. A period with a temperature low enough to initiate flowering and to check further growth. The actual temperature requirement is a cultivar characteristic.
3. A warm period (relatively dry) for stem elongation and the development of flowers and seeds.

A simple but effective method of seed production is to select typical plants from the root or vegetable crop, dig them up at the end of the growing season, keep them in a cool store over winter, and plant them out again in spring. They then rapidly flower, and seed can be harvested in the hot days of summer.

If a crop is to be used solely for the production of seed, seed is planted late in the season at close spacing. The small plants that develop are left in the ground over winter and set seed in the following season. Alternatively, at high latitudes where there is a risk of severe frost, the plants may be lifted, stored over winter and replanted. Because of the low-temperature requirement, seed production in the tropics is restricted to high altitudes.

Forage crops

Grasses and leguminous forage crops are mostly perennials, though some are biennials or annuals. Multiplication of their seed has special problems.

The seedlings grow very slowly, giving no return to the farmer in the first season, but providing ample opportunity for the smothering growth of weeds. Some farmers, therefore, sow a short-lived annual crop on the same ground; this is called a "cover" or a "nurse" crop. The nurse crop has an upright rather than a spreading habit of growth, and is sown thinly, or in rows wider than normal, so that it occupies the space that might become filled with weeds, but without smothering the seedlings of the main crop. When the nurse crop is harvested, the ground is left free for the growth of the seed crop, and fertilizer should be applied at this time to encourage it. Crop species used for this purpose are barley, yielding a grain harvest, and rape, providing grazing for sheep. It should be realized, however, that a nurse crop does check the growth of a seed crop to some degree, and the system is not applicable to all species. It depends on the ability to recover after the nurse crop harvest and to produce fertile tillers in the following spring.

In a grass seed crop, selective herbicides can be used to destroy dicotyledonous weeds, but it is more difficult to control grass weeds in this way. Sowing in rows permits inter-row cultivation, but grass seedlings grow so slowly as to be invisible in an early flush of weeds. Some seeds of a vigorous but shortlived species, such as mustard, are sometimes mixed with the grass seeds to mark the rows by their quicker growth and facilitate early cultivation.

Grasses of the temperate regions have a cold requirement for flowering, and flowering shoots are produced in the second year only by tillers which were initiated in the first season and have experienced the cold of winter. It is important, therefore, to encourage good growth and the production of numerous tillers in the first season, and for many species this involves early sowing.

A seed crop is sometimes grazed in the early part of the season. This may be done to delay flowering and seed production until the weather is favourable for ripening and harvesting, or to stimulate lateral growth and the development of more flowers. Grazing after harvest may be useful in removing old growth and at the same time providing valuable feeding for livestock. However, grazing can only be practised with species or cultivars which are capable of forming flower buds in the temperature and day-length conditions prevailing after grazing.

More than one harvest is possible, but certification schemes limit the number of harvests that can be taken from the same crop. Some seeds are shed on to the ground and germinate between the rows, and the resultant plants produce seed which is harvested along with seed from the original plants. Some of the seed harvested is, therefore, liable to be of a later generation and subject to genetic shift. In the tropics two harvests per year may be possible, e.g. *Chloris gayana* in Kenya.

The seed yields of grass crops are low because every tiller does not bear an inflorescence and every floret does not produce a seed, and the actual yield may be less than 30 percent of the theoretical potential. Future improvements in temperate regions are likely to develop from the introduction of new husbandry practices aimed at increasing the number of seeds which develop and ripen on each plant.

In tropical grasses there seems to be no antagonism between seed yield and herbage yield. Vigour, earliness, seed development, drought tolerance and persistence are compatible, and early-flowering types are favoured. Prolonged flowering gives rise to seed shedding and immaturity, so actual harvest yields are very low compared with temperate grasses but, on the other hand, seed rates are low and little seed is required. Because many of the seeds are empty and non-viable, even after cleaning, yields are expressed in terms of pure germinating seed (PGS). Seed is sown in narrow rows (300–450 mm) in order to restrict tillering and thus make flowering more uniform. Rapid growth is encouraged by early application of nitrogen.

In some tropical grass species, sterile hybrids and chromosome irregularities are common and very little seed is set. Useful grasses in the genera *Pennisetum* and *Cynodon* may therefore have to be propagated vegetatively by rhizomes or tillers. Vegetative planting needs much labour compared with sowing seed, so plant breeders are developing cultivars which produce abundant seed.

FURTHER READING

Bruinsma, J. (1966) "Analysis of growth, development and yield in a spacing experiment with winter rye," *Netherlands Journal of Agricultural Science*, Vol. 14, p. 198.

Department of Agriculture (1961) *Yearbook of Agriculture*, Washington.

Griffiths, D. J. and Others (1967) "Principles of herbage seed production," Welsh Plant Breeding Station Technical Bulletin No. 1, "The seed yield potential of grasses," Welsh Plant Breeding Station Report for 1973, p. 117.

FAO (1961) *Agricultural and Horticultural Seeds*, chapter 4, Rome.

Holliday, R. (1960) "Plant population and crop yield," *Field Crop Abstracts*, Vol. 13, No. 3.

Humphreys, L. R. (1975) *Tropical Pasture Seed Production*, FAO, Rome, (AGP:PFC/20).

Ministry of Agriculture, Fisheries and Food (1968) "Grass and clover crops for seed," Bulletin No. 204, London.

7

HARVESTING AND DRYING

WHEN THE DEVELOPMENT OF THE EMBRYO AND THE ACCUMULATION OF storage material is near completion, the seed has a moisture content of more than 50 percent[1]. The aim of the farmer is to have this seed in store a few weeks later with a moisture content of less than 15 percent. At the beginning of this drying process, the seed is attached to the mother plant, and drying is promoted by the natural agency of sun and wind. Then the farmer intervenes by starting his harvesting operations, and subsequent drying may be by natural or artificial means, depending on the system. Nevertheless, this loss of moisture should be regarded as a single process—it may be fast or slow, it may even go into reverse, but it determines the ultimate quality of the seed produced.

As pre-harvest drying depends on weather only, it may seem to be beyond man's control, but this is not altogether true. Seed production is practised in that part of a country which provides the best weather at the time of seed ripening and harvest. It may be possible to carry this concept a stage further and change the date of sowing so that ripening and harvest operations coincide with dry weather. This might require supplemental irrigation, or it might involve a serious reduction in yield, or it might even be quite impossible for rain or day-length adapted cultivars, but it is worth considering in planning

[1] It is customary to express the moisture content of seeds as a percentage of the actual weight of the seeds, i.e. as a percentage of the "wet weight".

seed production projects. Cutting or grazing in the early part of the season delays flowering, and it is possible in this way to manipulate the harvest dates of biennial and perennial forage crops.

Harvesting of most crops comprises two operations—cutting off parts of the plants, followed by separation of the seeds from the cut material. This second operation is called *threshing*. Each of these operations may be performed manually or by machine; a machine capable of carrying out both operations in one pass over the crop is known as a *combine-harvester*. Mechanized methods are not necessarily more effective than manual methods, and the latter may be preferred for economic or social reasons. The greatest advantage of mechanization, apart from economy of labour, is the power to deal with a large area in a brief spell of dry weather.

After these harvesting operations, depending on the moisture content achieved, a final drying of the seed by artificial means may be necessary.

Harvesting systems

Seeds of some crop species are harvested by hand picking. Workers walk through the crop picking off ripe seeds or fruits and carrying them in baskets. This practice is followed in small areas of vegetable crops and for species in which all the seed does not ripen simultaneously, e.g. cotton, pigeon pea, carrot, and some cultivars of *Pennisetum*. The process has to be repeated two or three times as more seed ripens, but for the final picking the plants may be cut and threshed. In maize, entire ears are harvested, and in some countries this is commonly done by hand.

Apart from hand picking, there are two common harvesting systems—cutting and threshing as two separate operations, and as one combined operation. Whether one or other system is better in practice depends on three factors: the effectiveness of natural drying, the losses sustained by shedding, and the extent of mechanical damage to the seed. Mechanical damage may be due to faulty setting of the threshing mechanism, but the nature and extent of the damage is influenced by the moisture content of the seed. If the seeds are too wet, they are crushed and the tissues bruised; if too dry they are liable to be fractured.

Cutting and threshing

The usual procedure is for the plants to be cut, left in the field to dry for a time, and then threshed. The cutting is normally done with a machine but, in some parts of the world, hand cutting with a scythe, sickle or knife is still practised. At the time of cutting the seed is still too wet to separate readily from the mother plant and is very susceptible to mechanical damage. Threshing follows when the seed has dried sufficiently.

In some crop species, the stem and leaf tissues gradually turn yellow, and by the time the seeds are ripe these tissues are dead, or nearly so. The plants are cut before the seeds have begun to shed, and are tied into bundles (or sheaves). After a period of natural drying, the sheaves are carried to a nearby threshing floor or static machine for immediate threshing, or they may be stacked for a time before threshing. If there is a risk of shedding, losses can be reduced by cutting in the early morning while humidity is still high or, more effectively, by fitting trays under the cutting machine to catch the shed seed. This is the traditional method for small-grain cereals, for some pulses, and can be used for erect growing grasses.

In other crop species while the seeds ripen the stems and leaves remain green, and a mass of vegetative matter has to be wilted by exposure to the sun and the wind. In these crops the cut plants are left lying in a windrow, or in small heaps, or they may even be hung on tripods or racks in wet situations. When dry enough for threshing, they are either carried to a static threshing machine, or picked up from the windrow and threshed by a moving combine-harvester. This is the normal method for herbage crops and many pulses, and can be used as a fall-back method for any crop in unfavourable weather. The crop should be cut as high as possible, partly to minimize the amount of stem and leaf that has to be dried and fed into the thresher, and partly to expedite drying by leaving a long stubble to keep the windrow off the ground. To avoid shaking ripe seeds on to the ground, the material should not be disturbed before threshing.

The success of the windrow method in avoiding losses by shedding is well illustrated by the results of an experiment in Nebraska, USA, with sweet clover. Windrowing followed by threshing was compared with direct combining 12 and 41 days later. The yields of threshed seed per hectare were:

Windrowed and then threshed	297 kg
Combined direct 12 days later	88 kg
Combined direct 41 days later	20 kg

A variant of this method is to cut the crop and carry all the material to a drier; threshing follows after drying. This technique has been used for peas and for the tropical grass *Setaria anceps*. Other variants are used for maize and groundnuts.

Maize has its own peculiar problems. Corresponding to the cutting and threshing of other crops are the two operations of *picking* and *shelling*. Picking is the removal of the ears from the plants, and shelling is the separation of the kernels from the ear. If left on the plant too long in warm or humid conditions, the ears are liable to be attacked by weevils and moulds, so picking is done early while the moisture is still quite high. Shelling at this stage, however, would damage the kernels, and this operation is carried out later when the ears have dried. For grain production, picking and shelling can be done at the same time, an operation that corresponds to combine-harvesting of small-grain cereals, but this causes too much damage to the kernels to be acceptable for seed production.

For seed production, the ears are picked by hand or machine, and some of the husks may be left to provide protection. With mechanical picking, damage is likely to be severe at moisture contents greater than 30 percent, and as much as possible should be picked at moisture contents between 20 and 25 percent. The ears are allowed to dry in open-sided cribs (figure 7.1) or in heaps, depending on the climate, to a moisture content of 12 to 14 percent, when they can be shelled. Alternatively, the ears can be dried artificially after removal of the husks.

Shelling is done either by hand or mechanically but, before this operation, the husks are removed and the ears examined. This provides an opportunity to remove any faulty ears—ears that have been attacked by moulds or damaged by insects, and ears that are not true to the cultivar type in number of rows, or shape and colour of the kernels. This, in effect, is a supplement to the roguing operation carried out earlier in the field.

Another crop with special problems is groundnut, which produces its seeds in underground pods. The whole plants are dug up manually, or by machine, and left on the ground to dry in windrows or small heaps. The pods are separated from the plants either in a static thresher or by a combine-harvester picking up from the windrows. Ripening is very uneven. In some cultivars, flowering extends over two or three months, and ripening over a similar period. There is, therefore, a mixture of ripe and immature seeds at harvest time, and there may even be differences in maturity within the same pod. Early harvesting produces many empty pods, and seeds that are immature and shrivelled. If the harvest is late, some of the early ripened seeds (if it is not a dormant cultivar) may have swollen and started to germinate. This swelling may cause splitting of the pods, thus admitting soil fungi which have free access to the seeds through the ruptured seedcoats. The inside of a pod, like the space enclosed by the husks of maize, provides a suitable environment for the growth of fungi both before and after harvest.

Figure 7.1 Maize cribs in East Africa (courtesy R. G. Griffiths).

Direct combine-harvesting

In this system the machine cuts off the inflorescences with a variable amount of stem and leaf and threshes them direct. This is now the normal method for harvesting cereals on a medium to large scale, but can be used for any crop provided the seed is retained on the standing plant until it is all dry enough for threshing, and that there is not too much vegetative material to be handled by the machine.

In certain cases it is possible to accelerate the drying of standing plants by spraying with a chemical agent. Except in very wet weather they dry sufficiently in anything from 3 to 10 days to be combine-harvested. This method has been used successfully for pulses and lucerne, and for some grasses and clovers. Three types of chemical have been used, each functioning in a different way:

1. Dessicants, e.g. diquat and paraquat. These substances damage the cell membranes, and death of the green tissues follows.
2. Defoliants, e.g. ethephon. This causes the release within the leaf of ethylene, a substance which specifically causes abscission.
3. Soil sterilants which damage the roots and stop water intake.

The most effective agents seem to be the desiccants, but they are costly to use, and there is a risk of discoloration and poor germinability of the seed.

An interesting technique (which has been tried on an experimental scale) is to spray the inflorescences shortly before maturity with a water-soluble plastic resin. This prevents physical shedding of seeds as they ripen, and harvesting can be delayed until all the seeds are fully ripe.

Tropical pasture plants

In many species of tropical pasture plants the seeds ripen over a long period and successively fall to the ground. Conventional harvest methods secure only the ripe or ripening seeds that are on the crop at the time of harvest, and this may be only a small proportion of the total seed production over the season. To overcome this problem, two new harvesting methods have been used with some success. In one of these methods a machine moves through the crop beating or rubbing the inflorescences to detach the ripe seeds, leaving the unripe ones to ripen on the plant and be collected later. This technique is applicable, for example, to the grasses *Cenchrus ciliaris* and *Paspalum dilatatum*. The other method is to allow the seed to fall to the ground and pick it up with a suction harvester. Seed of the leguminous crop *Stylosanthes humulis* has been harvested in this way.

Moisture content

In practice the most precise indication of a crop's fitness for cutting or threshing is often the moisture content of the seed. The available information for a number of crops can be summarized as follows:

Small grain cereals	Mechanical damage is minimized when threshed or combined within the following limits: Wheat 16–19 percent Barley 17–23 percent Oats 19–21 percent Rice 17–23 percent
Maize	Ideally, pick at 20–25 percent, but some of the crop may be picked at up to 30 percent. Ears from pollen plants are picked earlier at 30–35 percent.
Soya	Harvest at not more than 14 percent. Very sensitive to damage at moisture content above 15 percent or below 13 percent.
Groundnut	Shell at 10 percent. Very sensitive to damage.
Phaseolus and pea	Harvest at 14–20 percent.
Flax and Linseed	Cut at not more than 35 percent.
Ryegrass	Cut at 40–50 percent. Thresh or combine at not more than 35 percent. Depends on cultivar.
Dactylis	Cut at less than 40 percent. Thresh at 30 percent or less. Liable to shed seed if combined direct.

Electric moisture meters are unreliable at high moisture contents and may therefore be unsuitable for determining when a crop is dry enough for harvest. This applies particularly to grass crops in which moisture contents over 40 percent may be found shortly before cutting. For such conditions the infra-red method of moisture determination is advisable. It is not suggested that every seed grower should have a moisture meter. There are other indications of a crop's readiness for harvest, such as hardness and colour, and these are understood by experienced farmers. Nevertheless, for people who have to make decisions as to the harvesting of large areas or who have to advise farmers, these instruments are invaluable.

Whatever the local policy may be as to the best stage at which to harvest seed, in the event the decisive factor may be the weather or the availability of harvesting machinery, but even in perfect conditions of weather and management the decision to proceed may not be an easy one to make.

Threshing methods

The basic principles of threshing are as old as farming itself. The operation comprises three steps.

1. Beat or rub the plant material to detach the seeds.
2. Winnow to remove chaff, straw and other light material from the seed.
3. Sieve to remove heavier material, such as stones, soil and seeds of a different size.

Traditionally, the operation is carried out on a smooth floor. Separation of the seeds is achieved by manually beating the stems and inflorescences with a stick (called a *flail*), trampling by animals or men, or by driving a rubber-tyred vehicle over the material. In another method, a sledge on a set of metal discs like a disc-harrow is drawn by animals over the material. Winnowing is by wind, supplemented by the ingenious use of flexible trays or plaited bamboo sheets. Finally, hand sieves are used to separate heavy particles smaller and larger than the seed. It is a slow and wasteful process—much good seed is lost through the depredations of birds and vermin during the operation, and quality may be affected by the onset of rain before it is finished. The end product is far from pure, but it can be improved by hand picking. When the seed is for the farmer's own use and the quantity is small, selected ears may be threshed out entirely by hand.

Mechanical threshers are driven by oil, steam or water power, but the process is essentially the same. Beaters, corresponding to flails, revolve at high speed within a drum; the plant material is fed into this drum and beaten against its concave surface to detach the seeds. The material is then subjected to a strong air blast and shaken over sieves of different sizes. The product that emerges is fairly, but not absolutely, pure.

The components of a thresher (if operated separately) may each be within the power of a man. For cereals, threshing drums and fans of simple construction driven manually by a pedal or handle are sometimes used. The drums work better for rice, with its easily detachable seeds, than for wheat and barley.

The various mechanisms are adjustable so that the machine can be adapted to different crop species and to different conditions of the seed. The main consideration is how to produce a clean product, but without losing too much seed or inflicting too much damage. Susceptibility to mechanical damage depends partly on the kind of seed, and partly on its moisture content. Oily seeds, such as groundnut, sesame and linseed, are more susceptible than cereals. Wet seed tends to be crushed or skinned, and dry seed to be broken, but the material fed into a static thresher is usually well dried. To work efficiently the machine must be exactly level, and this is checked with a spirit level. The most important adjustments to be made are in the speed of the beaters, the width of the gap between the beaters and the concave, the air flow and the sieve sizes. The output should be kept under observation, so that adjustments can be made when necessary.

The thresher is a frequent source of seed contamination. At the end of a threshing operation, seeds are liable to be left inside the machine and introduced into the next lot, which may be a different species or, more insidiously, a different cultivar of the same species. Certain cleaning procedures must therefore be followed as a matter of routine.

At the end of an operation the machine is run empty until no more seed comes through. The operator then opens up the machine and cleans it out thoroughly, probing into all the most inaccessible places. It is a tedious task, taking up to 5 or 6 hours, but must be insisted on; vacuum cleaners and compressed air blowers are helpful. When there is a change from one cultivar to another, discard the first few bags of the second cultivar as they may be contaminated despite previous cleaning. In changing from a lower to a higher grade of the same cultivar, downgrade the first few bags. The sacks or other containers into which the threshed seed flows must be clean, as sacks previously used for seed are liable to be contaminated. It is important that the identity of each lot passing through the machine be retained, and that the containers in which the seed is removed be correctly labelled.

In a combine-harvester, the plant material is elevated from the knife blades into the threshing mechanism. In addition to the adjustments that are necessary in a static machine, adjustments have to be made in cutting height and forward speed and in the pick-up reel. If the machine moves too fast, the amount of material it takes in exceeds the capacity of the threshing mechanism; the seed is not completely separated, and some is damaged. The knives and the pick-up reel have to be adjusted upwards or downwards according to the height and bulk of the crop. In pulses, some of the pods may be borne low on the stem, and either damaged or left on the stubble if the knife is set too high. If the reel is revolving too fast, pods of pulse crops may be shattered and the seed lost. The moisture content of the seeds is liabile to be greater than for a stationary machine, and this necessitates a lower beater speed to prevent crushing; if the speed is too slow, however, the seeds are not all detached.

Compared with a static machine, the material that a combine-harvester has to handle is much more variable. The crop may not be uniform, being tall or short, standing or lodged in different parts of the field. Furthermore, the moisture content is liable to vary, not only from day to day, but during the same day; in Britain it can change by 3 percent in a day, even if no rain falls. The operator must, therefore, keep watch on the threshed seed (looking for damaged seed) and on the straw and other trash that is left on the ground (looking for unthreshed seed), and be prepared to make frequent adjustments. To discourage fast working, it is advisable to pay combine contractors by the hour.

A seed company may organize a service of mobile threshers. Under such a scheme, a machine visits a large farm or a central point convenient for a number of small farms. The company provides bags, labels and transport so that the threshed seed can be removed immediately to a store or processing plant.

Drying

Depending on the climate and the harvesting method followed, seed may or may not be dry enough for storage after the threshing process. In a dry climate, seed can go direct from the combine-harvester into store, but in less favourable circumstances further drying may be necessary. When seed comes off the thresher or combine-harvester, if it is not already dry enough, it must be dried down to within 3 percent of the moisture content required for storage within 24 hours. A second drying may follow later. At this stage the seed contains various impurities, mainly chaff, pieces of stem and leaf, and weed seeds. Removal of these impurities facilitates drying, and it may be advisable to pass the seed through a simple pre-cleaner before drying it artificially. This cleaning is only partial; the main cleaning operation follows later after the seed has been dried and kept in store for some time.

The moisture content limit at this stage (i.e. 3 percent above storage requirement) is absolute and must be taken into account in planning any seed multiplication project. In regions of low rainfall, drying may not be necessary, or blowing air heated to about 5°C above the ambient temperature may be sufficient; simple driers for this purpose are often installed on medium-sized to large farms. In wet regions, particularly in the tropics, large-scale drying plants are necessary. To avoid delays, drying units need to be located in each seed-producing district and provided with mechanical transport to bring seed in rapidly from outlying farms. During the harvest season, they need to operate 24 hours a day.

It is characteristic of agricultural seeds that they can withstand the removal of water and remain viable even if the moisture content is reduced as low as 5 percent. In contrast, the seeds of some tropical rain forest species lose their vitality if the moisture content is reduced below 15 percent, and storage is impossible.

Nevertheless, artificial drying can depress the germinability of agricultural seed, giving rise to abnormal seedlings, affecting the permeability of the seedcoat, destroying enzymes, or causing the outer layers to become hard so that, when the embryo subsequently imbibes water and swells, fractures and cracks develop. In pulses, the testa cracks and separates from the cotyledons, and the seedlings that emerge show reduced vigour and poor establishment. Why these defects develop is not understood. In some cases it seems to depend on the temperature of the hot air, and the time during which the seed is exposed to it. In other cases it seems to be caused by rapid drying, even if the temperature is low, e.g. soya seed dried by air with a relative humidity of less than 40 loses viability even at low temperatures. In immature seed the moisture is deep-seated and difficult to remove, but in ripe seed which has attained the same moisture content by wetting with rain, the moisture is superficial and easier to drive off. To minimize these adverse effects, it is customary to dry rice in two stages.

In general, temperatures up to 45°C are safe, but higher temperatures may be used in continuous-flow driers than in batch driers, because the time of exposure is shorter. It is not possible to give maximum drying temperatures which are guaranteed to be safe, but the following seed temperatures provide a guide. When two temperatures are given, the higher is applicable only to relatively dry seed in a continuous-flow drier; otherwise the lower temperature is recommended.

Wheat, barley, oats	43–65°C
Rice	50–60°C
Maize	40–45°C
Pea	30–50°C
Groundnut	36°C
Beet	45°C
Soya	30–55°C
Ryegrass	40–70°C
Clovers and brassicas	27–40°C
Onion	21–33°C

It is important that after drying the seed be cooled by forced ventilation before going into store.

Relative humidity

A seed can be regarded as a structure composed of complex substances such as cellulose, starch, oil and protein, with some water in addition. This water is variabile in quantity—it can be added to or removed. If a seed is placed in water, it will imbibe some, thus increasing its moisture content; if it is then removed from the water, the seed will dry, i.e. water will be removed and the moisture content will fall.

The water molecules in a seed are always in a state of vibration and, if near the surface, occasionally escape into the air; this is *evaporation*. Similarly, the water molecules in the air are vibrating; there may be fewer of them, but they move more violently, and some of them come in contact with the seed and enter into it; this is *absorption*. These two processes, evaporation and absorption, are going on continuously. When evaporation exceeds absorption, the seed dries out; when absorption exceeds evaporation, the seed takes up moisture from the air. In an intermediate condition, evaporation and absorption are equal; the moisture content of the seed is then said to be in equilibrium with the humidity of the air. When water evaporates, it changes from the liquid state to a vapour, but the number of freely-vibrating water molecules that can be contained in a given volume of air is limited. When air is carrying this maximum amount of water vapour, it is said to be *saturated*. The hotter the air, the more water it can carry, as shown in Table 7.1. A kilogram of air at a temperature of 10°C can carry only 7.6 grams of water, but at 40°C it can carry 41.4 grams.

Table 7.1 Maximum amount of water that can be held by air at different temperatures (from Harrington, J. F. in *Seed Ecology*, ed. Heydecker, 1973)

Temperature, °C	*Mass of water, grams per kilogram of air*
0	3.9
10	7.6
20	14.8
30	26.4
40	41.4

The water content of air is expressed as *relative humidity* (RH), which is the amount of water calculated as a percentage of the amount the same air can carry when saturated. At saturation point the relative humidity is 100 and, as the water-carrying capacity of air varies with temperature, so does relative humidity. For example, if air at a temperature of 20°C has a relative humidity of 90, it contains

13.3 g of water per kg (90 percent of 14.8). If this air is cooled to 10°C, the amount of water remains the same, but it is now in excess of the air's holding capacity of 7.6 g per kg and the excess water appears in the liquid state, i.e. some of the water condenses. Similarly, if air is heated, its water content remains the same; but its moisture-carrying capacity increases, and its relative humidity falls.

When the moisture content of the seed is in equilibrium with the relative humidity of the adjacent air, no change in the moisture content of the seed occurs. For drying to take place, the air must have a relative humidity below this equilibrium point, and the process can be accelerated by blowing air through the mass. However, if the relative humidity of the air is close to or above the equilibrium point, it must be heated to increase its water-carrying capacity and lower its relative humidity.

Traditional methods

In every country where drying is necessary, farmers have, on the basis of experience, evolved methods which utilize the sun and wind, and which are suitable for small quantities. Threshed seed may be spread out in a thin layer on a smooth earthen floor or on straw matting. Ventilation is improved by stretching the matting on a horizontal framework supported on stakes above ground level to allow the wind to blow through the seeds. Unthreshed inflorescences may be hung on frames or placed in cribs with open sides. As germination capacity may be affected by direct sunlight, due to the high temperature if the moisture content is high and perhaps to ultra-violet radiation, prolonged exposure should be avoided.

Batch driers

In a batch drier, relatively dry air is blown through a layer of seed until drying is completed. The seed is then removed and replaced by another batch. The method is simple, suited to small quantities, allows easy cleaning, and is recommended for farm drying.

In a horizontal drier, the seed is contained in a box or chamber with a perforated or slatted floor through which the air is blown. Air ducts can be laid on a barn floor and the seed piled over them. Another modification is the sack drier, in which the seed, contained in a woven sack, is placed on top of a grid from which air is blown (figure 7.2). A cylindrical storage bin with a raised perforated floor can be used and the air blown underneath the floor (figure 7.3). A vertical drier consists of two concentric cylinders, both being perforated (figure 7.4). The space between the two cylinders is filled

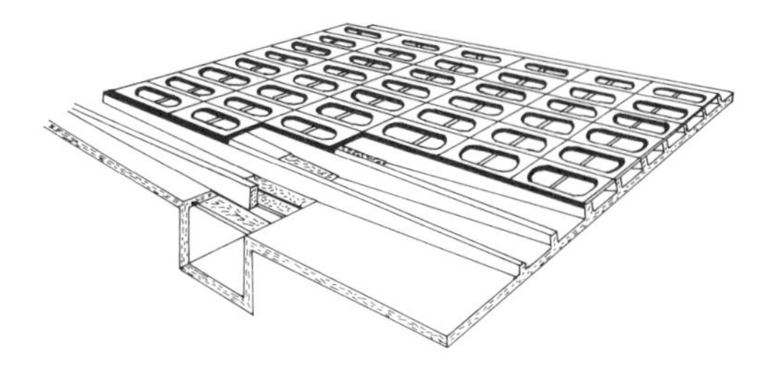

Figure 7.2 Sack drier (redrawn from *Proc. ISTA*, Vol. 28, 1963).

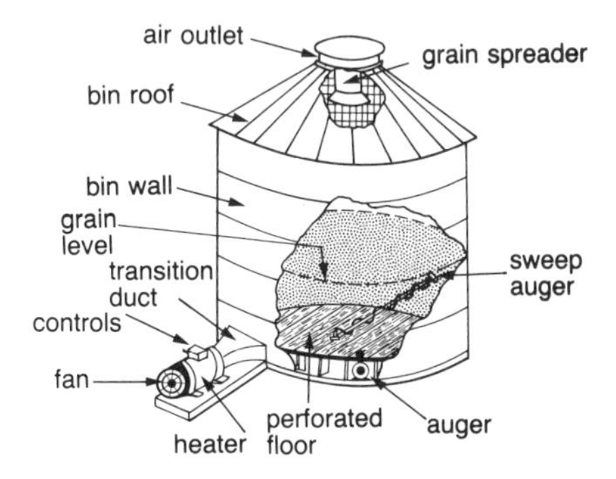

Figure 7.3 Bin drier (redrawn from *Proc. ISTA*, Vol. 28, 1963).

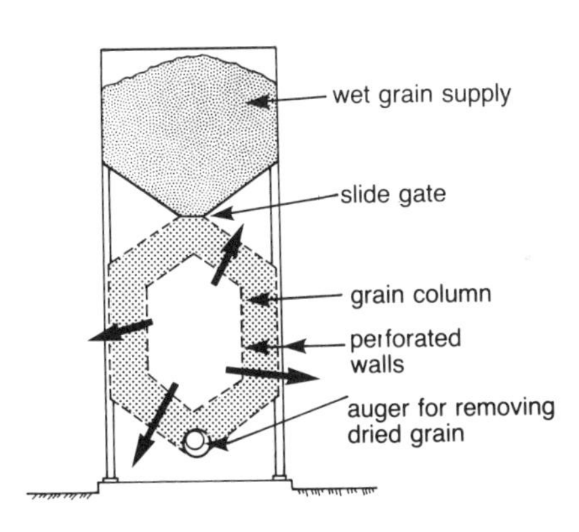

Figure 7.4 Vertical drier (redrawn from *Proc. ISTA*, Vol. 28, 1963).

with seed, and air is blown into the inner one, from where it passes outwards through the seed. Drying rates depend on the size of the batch.

A drier need not be an elaborate affair. A small batch drier can be constructed from locally available materials, including the ubiquitous oil drum, and fuelled with wood. Driers of this kind can be effective, but are not dependable; they lack temperature control and have to be continuously watched.

In the horizontal drier the seed at the bottom is dried first, and the dry zone gradually extends upwards. If the layer is too thick or the air flow inadequate, drying of the uppermost seed may be unduly delayed. The layer should not exceed a depth of 3 metres, but for seed of high moisture content this should be reduced to one metre, or less for forage seeds. If the drying is done in a storage bin, a layer of undried seed may be added on top of the dried batch and the process continued, but only if the seed is already fairly dry and the air passing through is not too hot. The efficiency of this process can be improved by drying in two stages. When the first batch is partially dry, the emerging air is passed through a second batch held in another chamber; the process is then repeated with the second and a third batch, and so on.

The air that is blown through a batch must not be too hot, because the seed at the bottom will be kept at the temperature of the entering air for some time and may be overheated. Indeed, it may not be necessary to heat the air at all. Heating to less than 10°C above the ambient temperature can be very effective, but on a hot humid day in the tropics, even a few degrees above ambient temperature may be high enough to harm the seeds. In these circumstances, dehumidification of the air is necessary.

As the air rises through the seed mass, it cools, and its capacity to hold water decreases. If hot air is blown through, it picks up much water from the lower levels and, when it reaches the upper surface, it may cool down so much as to become saturated and deposit liquid water. The result is that moisture is transferred from the bottom to the top, where it may become excessive for a time. It may seem a paradox, but it follows that the wetter the seed, the cooler the blown air should be. The temperature of the air can, of course, be raised as the drying proceeds.

The rate of drying is important; if it is either too rapid or too slow, the seed is liable to be damaged. Rapid drying can harm the seed, either because water is withdrawn too quickly or because of the high temperature. Slow drying may have the effect of maintaining the seed at a high moisture content and a relatively high temperature, and so accelerate the deterioration which drying is intended to prevent.

Continuous-flow driers

In this type of drier the seed moves horizontally or vertically through a stream of hot air and then into a cooling chamber (figure 7.5). It is a continuous process on a factory scale and is therefore suitable for

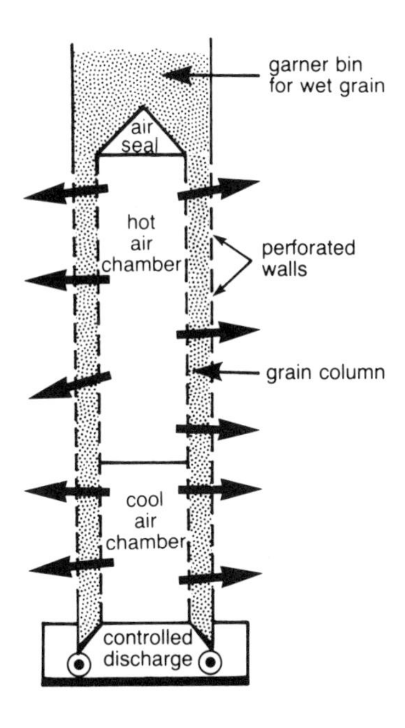

Figure 7.5 Continuous-flow drier (redrawn from *Proc. ISTA*, Vol. 28, 1963).

large quantities. It is, however, difficult to clean when there is a change of cultivar. Because the seed is heated for a much shorter time, the air temperature can be higher than in a batch drier.

FURTHER READING

ISTA (1963) "Drying and Storage," *Proceedings International Seed Testing Association*, Vol. 28, No. 4.
—— "Seed Storage and Drying," *Seed Science and Technology*, Vol. 1, No. 3, 1973.

8

STORAGE

MOST OF THE WORLD'S SEED HAS TO BE CONSERVED FOR A PERIOD OF several months, from harvest until the next sowing season. When a farmer sows his own seed, it is stored on the farm, the quantity is small, and traditional methods usually suffice; but when the seed has to be certified and sold to other farmers, the quantities are very much larger, and conservation becomes more difficult. Some seed has to be carried over for a further year, requiring a storage period of 18 months or more. This applies to high-grade seed surplus to requirements, and reserve stocks held as a safeguard against years of shortage.

Storage for even longer periods is possible, but is expensive and can be justified only if the seed is of particular value and the quantities small. Examples are plant breeders' germ plasm stocks, and samples held by certification authorities as controls against which each season's multiplication can be checked for cultivar authenticity. Horticultural seeds are sold in relatively small quantities and, if there are difficulties in producing seed, e.g. because of special pollination requirements, it may be advantageous not to produce seed every year and to carry over stocks.

The aim in storage is to maintain the germination capacity of the seed, and generally this requires more stringent conditions than the conservation of nutritional or industrial qualities. After harvest, seed maintains its original germination capacity, or nearly so, for a period of weeks, months, or even years; it then degenerates steadily and maybe rapidly (figure 8.1). For example, a lot of pea seed was stored in poor conditions; its germination capacity of 93 percent persisted for 12 weeks, but fell to zero in the following 18 weeks. For practical purposes, storage life comes to an end when the germination capacity *begins* to fall.

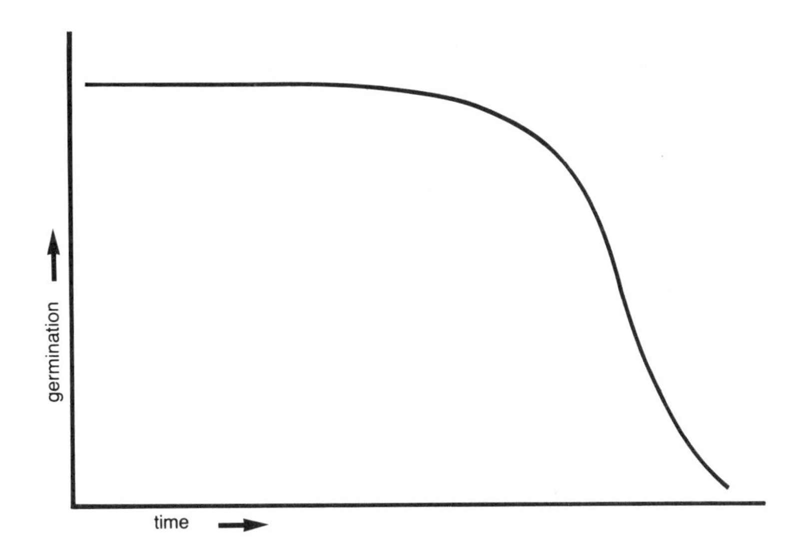

Figure 8.1 Typical graph showing how the germination capacity of a seed lot changes during storage.

A seed lot which has ripened under good conditions, has been well harvested, and has been properly dried, possesses, when it goes into storage, a high field planting value, i.e. the seeds are capable of producing strong seedlings able to establish themselves in adverse field conditions. During storage some of this value is lost, slowly or rapidly, depending on storage conditions. Germination capacity is defined as *the percentage of seeds capable of germinating to produce seedlings robust enough to establish in* good *field conditions*. Seed which still has a high germination capacity, therefore, may have lost some of its original field planting value. No numerical measure can be placed on field planting value; it is a concept combining germination capacity and seedling vigour.

During the phase when the original germination capacity is maintained in storage, some deterioration is taking place and can be demonstrated by measurements of respiration rate, enzyme activity, leaching of solutes and seedling growth rate. In the phase of rapidly falling germination capacity which follows, the embryos lose their ability to develop into normal seedlings and then die. In the example quoted above, this phase lasted 18 weeks, and in each week about five in every 100 seeds fell over the edge and lost their ability to germinate normally.

A store of seeds can be regarded as an ecological habitat corresponding, say, to meadow or forest. A meadow is inhabited not only by grasses, but by other plants, by animals, birds and insects, and the activity of any one species reacts on the others. In a store, in addition to seeds there may be fungi, bacteria, insects, mites, rodents and, from time to time, birds. Some kinds of insects and mites introduce or feed on fungi, or require the seed to be decomposed by fungi before they can eat it. Insect damage encourages fungal growth. Some fungi are poisonous to certain insects. One kind of insect may be antagonistic to another, or compatible so that they can multiply together. One activity that all these organisms have in common is respiration, which can be represented by the following chemical equation:

$$C_6H_{12}O_6 + 6O_2 \rightarrow 6CO_2 + 6H_2O + \text{heat}$$

The water and heat thus produced encourage the growth and development of all the organisms present, but some more than others perhaps, so that the balance between them alters. The heat of respiration can actually raise the temperature of a seed mass to above 60°C.

Rodents and birds

Loss due to rodents and birds can be severe, particularly in buildings of poor construction. In addition to the seed actually eaten, much is damaged or lost by spillage from torn sacks. Admixtures of cultivars can be caused by seeds being carried from one lot to another.

The best protection against rodents is a properly constructed store, built of stone, brick, concrete or metal, at least at the base. Small isolated stores should have a floor raised about a metre above ground, a rat-proof door, and a moveable entrance ramp. For a large building which is an integral part of a processing plant, a raised floor may not be possible, but the walls should have a smooth surface and projecting rodent guards to prevent climbing.

Insects and mites

Storage insects and mites are always present in the vicinity of human habitations and, in the tropics, flying insects may deposit their eggs on seeds in the field (figure 8.2). An initial contamination of seed put into storage cannot be entirely prevented; whether or not an original light infestation develops to a harmful extent, depends partly on the moisture content of the seeds and partly on the temperature. Most damage is caused by the immature stages of moths and beetles.

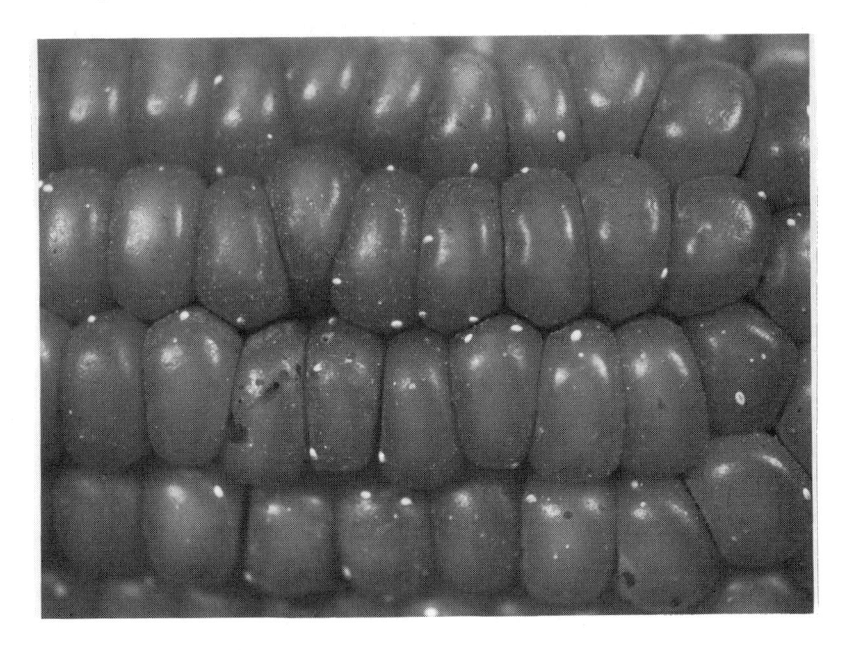

Figure 8.2 Eggs of bruchid beetles on maize (Ministry of Agriculture, Fisheries and Food).

Direct damage is due to feeding on the embryo and endosperm, and seeds that are already injured are preferred. Seeds with hard outer coverings are less susceptible than others, e.g. barley and oats compared with wheat and rye. Caterpillars of moths, such as the warehouse moth feed on many kinds of cereals and pulses, often preferring in the case of cereals the embryo to the endosperm. Many beetles, such as the granary weevil (figure 8.3), bore into the seed and lay their eggs internally; the larvae then feed on the endosperm and the embryo. In other species, the eggs are laid on the surface and the larvae bore into the seed (figure 8.4).

Damage to the embryo of an endospermic seed, or to the vital radicle/plumule axis of the embryo in a non-endospermic seed, can destroy the seed's germinability. If the damage is restricted to the endosperm or to the cotyledons, the embryo may be capable of

Figure 8.3 Granary weevils on wheat (Ministry of Agriculture, Fisheries and Food).

Figure 8.4 Lentils damaged by beetles. Egg capsules are on the surface and adults are ready to emerge (Ministry of Agriculture, Fisheries and Food).

growth, but its vigour is diminished in proportion to the amount of storage material destroyed.

Indirect damage follows from the insects' respiration. Enough water may be produced to induce germination of the seeds. As temperature rises, it accelerates the growth of the insects; it may eventually become lethal, but by that time fungi, which can withstand relatively high temperatures, have taken over and they induce further spoilage. If the infestation is not distributed throughout the bulk, heat develops at the infested spots, and convection air currents carry the moisture to cooler parts; there the moisture raises the humidity in the air spaces and encourages the growth of fungi. Webs and cocoons interfere with aeration.

Dry seeds are too hard for insects to feed on, and the most effective means of control is to dry down to a moisture content in the range 10–12 percent for cereals, or 6–8 percent for oil seeds, the actual percentage being determined by local experience. Sun drying has the additional effect of killing insect eggs laid on the surface. The optimum temperature for insect development is 25°C; there is little activity below 20°C and it practically stops below 10°C. Insect infestation is therefore a problem of the tropics. In cool temperate regions, infestation can be checked by blowing cool ambient air through the bulk.

Mites have a lower optimum temperature than insects, and consequently are more damaging in temperate regions. In general they feed on the endosperm, but some species prefer the embryo (figure 8.5).

Seed should not be kept in the vicinity of any food store which might contain infested grain and act as a continuing source of infection. The seed store should as far as possible be insect-proof, with smooth wall surfaces. Infestation can be controlled to some extent by minimizing carryover from lots previously held in the store. When empty between seasons, it should be thoroughly cleaned and either fumigated or have the walls sprayed with a persistent contact insecticide. The seed itself can be treated with insecticide. Fumigation can be carried out under sheets, or if an infestation develops during storage in a bin, a fumigant can be introduced into the air-flow and gently blown through the seed mass. Alternatively, contact insecticides in the form of liquid or pellets can be added to the seed as it passes into a sorage bin. Provided the moisture content and temperature are not too high, methyl bromide and phosphine can be used as fumigants, and malathion as a contact insecticide. However, it should be borne in mind that treatment with insecticides carries the risk of phytotoxic effects and loss of germination capacity. Practice should be based on specialist advice in the first place, and thereafter on experience.

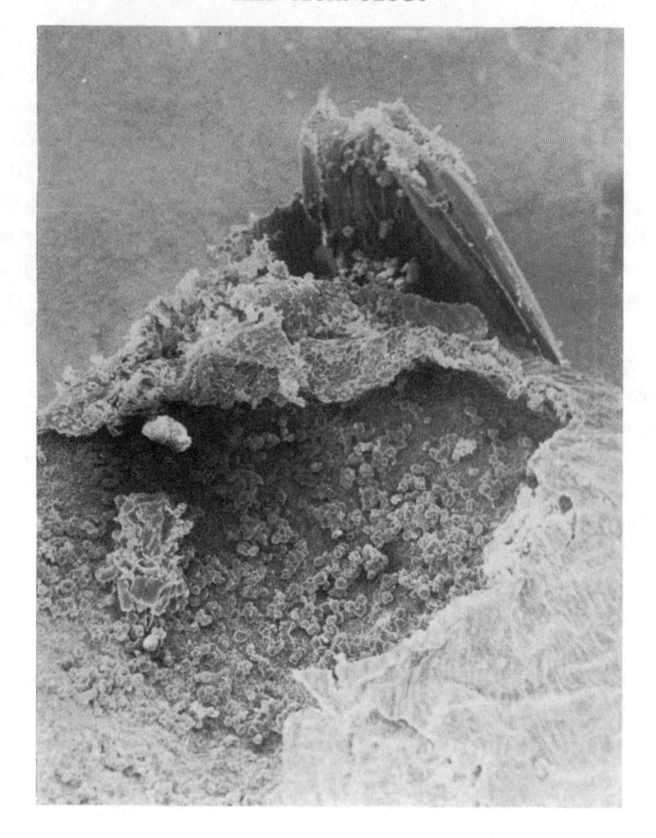

Figure 8.5 Barley grain damaged by mites (Institute of Terrestrial Ecology).

Fungi

In temperate regions, storage moulds, mostly belonging to the genera *Aspergillus* and *Penicillium*, are associated with severe spoilage. These fungi can grow on any kind of organic matter; they are universal and their spores are abundantly air-borne, as can be demonstrated if a sterilized agar plate is exposed. The spores infect the seeds superficially before, during and after harvest, particularly if wet weather has encouraged mould growth on dead plant tissue. It can be assumed, therefore, that every seed lot has been contaminated; to what extent the inoculum will grow, depends on the storage conditions.

The rate of growth of these fungi increases with temperature, and they can grow vigorously both above and below the temperature range best suited to insect development. This explains why they can take over when the seed lot becomes too hot for insects, and why they are more damaging than insects outside the tropics.

Their outstanding characteristic is an ability to grow in relatively dry situations, even in the absence of free water. In such conditions the fungi which parasitize plants in the field are incapable of growth, but the mould fungi can obtain their water from the vapour in the air. They can grow in a seed lot provided the relative humidity in the air spaces between the seeds is high enough. The fungal mycelium grows on the surface and in the superficial layers of the seeds.

Loss of seed viability is associated with fungal development, but the relationship may not be entirely a direct one. These fungi are not parasites and cannot invade living tissues. In the early stage of an infestation, fungal growth is on dead cells on the surface, in cracks and on damaged parts of the seed. Toxic substances are produced which on imbibition diffuse into the embryo and impair its viability. Subsequently, vigorous fungal growth produces water and raises the temperature; the physiological deterioration of the seed is then accelerated, and soluble substances can diffuse outwards to nourish the fungus. In extreme cases the seed may even germinate.

The key to the control of storage fungi is their humidity requirement. If the relative humidity in the surrounding air is less than 65–70 percent, depending on the species of fungus, they are unable to grow. Moulds can therefore be kept in check by maintaining the relative humidity of the air in the spaces between the seeds at or below this limit. Fungicides are ineffective because of the absence of liquid water. In any case, they involve the risk of phytotoxicity in prolonged storage.

The seed

Finally, what of the seed itself?

The physiological activity of the embryo is influenced by the amount of oxygen and water available, and by the temperature. For long storage life, this activity must be kept to a minimum.

Broken seed coats allow access to oxygen, and the respiration rate increases in consequence, but this by itself is not likely to be a critical factor in determining the storage life of large seed lots, nor is the oxygen content of the atmosphere in the store.

When the moisture content of the seeds exceeds about 30 percent, they may begin to germinate. The swelling of the embryo ruptures the seedcoat, and the products of respiration (water and heat) accelerate the process. This incipient germination promotes fungal activity and leads to the loss of germination capacity, even if the lot is subsequently dried, because dehydration of a germinating embryo is lethal.

In the absence of fungi and insects, seeds lose their vigour and viability, but more slowly. This deterioration is physiological, and its rate is influenced by moisture content and temperature. Respiration rate falls, enzyme activity is diminished, and the permeability of cell membranes is affected.

The effective storage life of seed depends on three factors:

the kind of seed
its pre-storage history (including harvesting and drying)
the actual conditions of storage.

Kind of seed

There is considerable variation in the length of time that seeds of different farm-crop species can survive in store, and in their ability to withstand poor storage conditions. Storability is a character that is innate in the species. The outstanding example is rice, which explains, in part at least, why it is the dominant crop in that part of the world where the climate is most inimical to seed storage. The millets have a long storage life and *Pennisetum*, for example, seems to be virtually immune to weevil damage. Other seeds that store well are oats, beet and lucerne. At the other extreme, seeds which are difficult to store are groundnut, soya and onion. None of these can be carried over a season without conditioned storage. In the case of groundnut and soya, this is probably linked to their extreme susceptibility to mechanical damage, which in turn follows from the fact that the testa is too thin to afford adequate protection. Groundnuts store better if shelling is postponed until after storage. Examples of seeds which are intermediate in this respect are maize, *Sorghum*, cotton, *Phaseolus* and clover.

Differences of this kind may even be found between cultivars of the same species. This is well known in maize, rice, wheat and lettuce; in soya it depends on the thickness of the testa. In sugar beet there appears to be a marked difference between monogerm and multigerm seed. Late cultivars may appear to store less well than early ones, because their seed is more likely to be exposed to inclement weather at harvest time and has therefore deteriorated to some extent before going into store.

This innate storability is only relative. When the condition of the seed is poor and the storage environment is adverse, no seed can maintain its viability for long.

Harvest effects

The condition of the seed when it is harvested has a decisive influence on its subsequent storability. Seeds harvested before maturity do not store well, and the crop should not be cut until they are well filled and ripened. Unsettled weather causes alternate wetting and drying of the seeds before and during harvest, maybe in warm weather; e.g. the moisture content of mature soya seeds can fluctuate between 11 and 20 percent at this time. In the wet tropics it is possible to dry seed naturally to a moisture content that is apparently suitable for storage. The drying process, however, is prolonged and erratic, and the seed does not store as well as seed dried quickly by artificial means to a similar moisture content. Another effect of wet weather is to increase the load of fungal spores that is carried into store and may develop later. In maize, in contrast to other cereals, the ear is enclosed within bracts, or husks, and this provides sheltered conditions for the development of fungi and insects which may multiply excessively if harvest is delayed.

Figure 8.6 illustrates dramatically the effect of season on storability. In an experiment at Cambridge, seed of the winter wheat cultivar Yeoman was put into near-perfect storage conditions each year after harvest. In each subsequent year, samples were drawn from the store and tested for germination capacity. The graph shows the changes in germination capacity that occurred in seed harvested in four contrasting years. The seed that kept best was harvested in 1921, a year of little rain and abundant sunshine. The worst seed was harvested in 1922, a year of high rainfall and cloud. The weather of 1926 was between these two extremes, and the seed turned out to be of intermediate keeping quality. The explanation of the poor storability of the 1929 seed is that because of drought the seed was exceptionally dry and liable to mechanical damage.

This effect of weather explains why seed from certain countries has a high reputation for keeping quality. These countries have a climate which provides consistently good harvest weather.

A common cause of loss of viability is mechanical damage incurred in harvesting or in subsequent processing. The damage may be an inevitable consequence of the process, e.g. shelling groundnuts, or harvesting maize with a picker-sheller. It may be due to faulty setting of the combine-harvester or to threshing at the wrong moisture content; in maize, 70 percent of seeds have been found damaged after shelling at the low moisture content of 8 percent. The effect of damage is to injure the embryo and facilitate the entry of fungi during storage.

Susceptibility to damage is to some extent influenced by size, shape and other characteristics. Of the cereals, rice, barley and oats are more resistant than wheat or rye, because the embryo is protected by the horny lemmas and pales. Round seeds are more resistant than flat ones, e.g. brassicas compared with sesame. Large seeds like pulses and

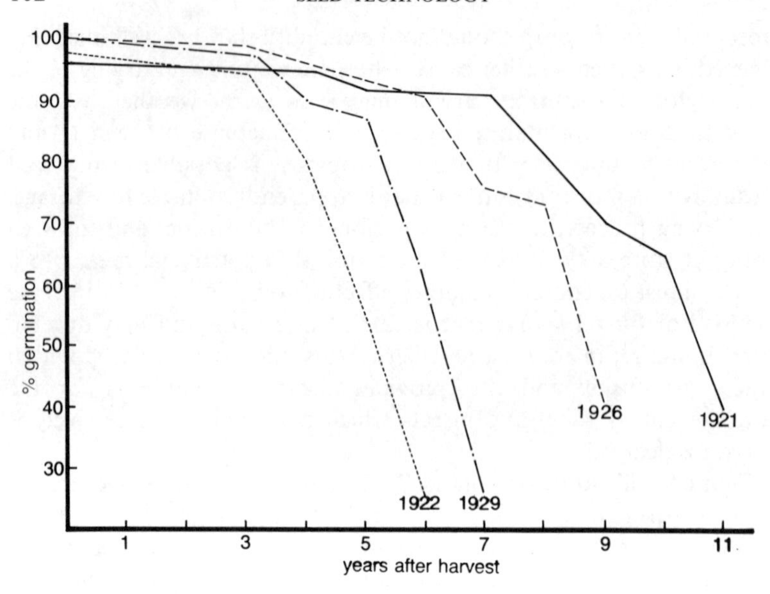

Figure 8.6 Loss of germination in wheat seed harvested in different years (compare with figure 8.1) (after *Journal NIAB*, Vol. 11, 1967).

seeds with a protruding radicle, such as sorghum, groundnut and some herbage legumes, are particularly liable to be damaged. Oil seeds are more susceptible than other seeds.

Successful storage depends to a great extent on the moisture content of the seed when it is put into store. Drying is an essential part of the harvesting process, and any failure to dry the seed down to a suitable moisture content is reflected in the seed's subsequent performance. The average moisture content may conceal considerable variation within the bulk. If not uniformly dried, the moisture content may be too high at isolated spots within the storage bin, and deterioration can start and spread from these points.

Finally, the seed must be well cleaned; pieces of leaf, stems and chaff tend to hold moisture more than seeds and to obstruct aeration in store.

Storage conditions

It has been explained in chapter 7 that water molecules are continuously moving from a seed into the adjacent air and vice versa, and that when the outward and inward movements are equal, the moisture content of the seed is said to be in equilibrium with the humidity of the air. When seeds are spread out in a thin layer in an open space, and so exposed to a large volume of air, the balance of movement may be outward or inward, but the moisture content of the seed will change until it is in equilibrium with the humidity of the air. In storage, the seeds are in a compact mass and a different situation exists. The surface of each seed is exposed only to the very small volume of air in a space between adjacent seeds. Assuming that there is no movement of air or water vapour within the seed mass, the humidity of the air in each space will change until it is in equilibrium with the moisture content of the surrounding seeds. If the initial relative humidity is higher than the equilibrium point, moisture will on balance be absorbed by the seeds; if it is below the equilibrium point, there will be a net movement of moisture from the seeds into the air space.

Actually, within a seed mass, unless there are significant differences in humidity or temperature, there is very little movement of air or water vapour. Furthermore, gaseous movement between the internal spaces and the ambient atmosphere of the storage building is very restricted. If the relative humidity of the ambient atmosphere differs from the relative humidity of the internal air spaces, the internal humidity will indeed change until both are the same but, except in excessively dry or humid conditions, the change is extremely slow. The moisture content of the seeds will, of course, change concomitantly.

Because of the relationship between the moisture content of seed and the humidity of the air, it is important to know the equilibrium points. The equilibrium moisture content of a seed depends on its chemical composition. Starch absorbs moisture readily and retains it tenaciously, and so tends to have a high moisture content. Oils, on the other hand, neither absorb nor retain moisture and tend to have a lower moisture content. Proteins are intermediate in this respect. Each crop species, therefore, tends to have its own typical moisture content for equilibrium, but the differences between species are liabile to be obscured by real or apparent variations within a species. Intra-specific differences may be due to cultivar characteristics, seed damage, temperature, or whether the seed is drying or wetting at the time. The main difference lies between starchy seeds, such as cereals and oil seeds; e.g. at RH 70 their moisture contents are round about 15 and 9 percent respectively (see Table 8.1 and figure 8.7).

The principal agents of seed deterioration in store are fungi and insects, and their development is influenced by temperature. Fungi, however, are primarily dependent on a high humidity in the interstitial spaces of the seed mass, and this in turn is dependent to a great degree on the moisture content of the seeds. The limiting relative humidity is between 65 and 70 percent, and the corresponding equilibrium moisture content is about 14 percent for starchy seeds and 8 percent for oily seeds. The multiplication of insects depends partly on the moisture content of the seed as influencing its texture,

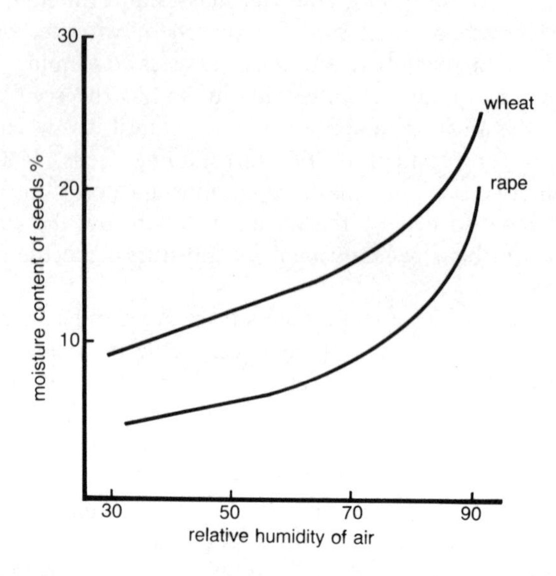

Figure 8.7 Curves showing how the equilibrium moisture content of wheat and rape seed changes with the relative humidity of the ambient air (based on Kreyger, J., IBVL, Wageningen, 1972).

Table 8.1 Typical moisture contents (percent) of seeds in equilibrium with a range of relative humidities from 30 to 70.

	Relative Humidity				
	30	40	50	60	70
Wheat, Rye	9	10	12	13	15
Barley, Oats	8	10	11	12	14
Maize	9	11	12	13	14
Sorghum	9	10	11	12	14
Rice	8	9	11	12	13
Ryegrass	9	10	11	12	14
Lucerne	8	9	10	12	14
Phaseolus	7	10	12	14	16
Pea	8	10	12	13	15
Beet, Onion	7	8	10	11	13
Soya	7	7	8	10	12
Cotton	6	7	8	9	10
Brassica	5	6	7	8	9
Groundnut	4	5	6	7	9

and partly on the humidity within the air spaces, but the limits are lower than for fungi. In general, they cannot develop at moisture contents below the equilibrium of RH 40, but there is considerable variation in this respect between different species.

If fungi and insects do not develop because moisture content, humidity and temperature are unfavourable (too dry, or too cold), seed will still deteriorate with age and lose viability. For large-scale storage of most kinds of farm seeds, however, deterioration due to ageing alone becomes significant only in long-term storage and does not affect season-to-season or reserve storage.

Taking all causes of deterioration into consideration, the main physical factors determining storage life are the moisture content and the temperature at which the seeds are maintained. Because of the part played by respiration, the availability of oxygen might be expected to have some influence. However, oxygen becomes limiting only in hermetically sealed storage, and can be ignored in the storage of large lots. Except in extreme conditions, the effects of moisture and temperature are summarized succinctly in "Harrington's rules", thus:

> for each one percent increase in moisture content, and for every 5°C rise in temperature, the storage life is halved.

The conditions necessary to maintain viability can be expressed simply enough as dry and cool, but the material and mechanical resources that are needed to provide this environment vary enormously according to climate. Another variable that has to be taken into account is the duration of storage required. Most seed has to be stored from harvest to seed-time only, but some stocks have to be stored for more than a year, and call for more stringent conditions. For this longer-term storage, a useful target to aim at is the temperature in degrees Fahrenheit, added to the relative humidity, not exceeding 100.

Because gaseous diffusion between the interstitial air spaces and the atmosphere of the store is very slow, and seed is a poor conductor of heat, a seed lot tends to maintain its original moisture content and temperature for a considerable time, longer or shorter according to the weather. The first requirement, therefore, is to have the seed well dried, and then cooled at the beginning of its storage period. Indeed, in dry and temperate regions, this may be all that is necessary.

In the temperature zone, deterioration is caused mainly by fungi. Therefore, seed has to be dried down to a moisture content in equilibrium with RH 65–70 or less (Table 8.1); cooling is simply a matter of blowing ambient air through the lot. Provided the seed is protected against rain, it will store over the winter months until the next sowing season. With experience it is possible to vary the actual moisture content according to the details of local circumstances, such as the kind of container, the structure and site of the warehouse, and the precise duration of storage. In the case of seed stored for longer periods, deterioration occurs mainly during the warm summer months. This can be controlled to some extent by aerating with cool air of low humidity, e.g. at night in dry weather, but more effective humidity control of the store may be necessary.

In the hot dry tropics and sub-tropics, insects are the principal cause of deterioration. Apart from providing shade, temperature control of large quantities is virtually impossible, and the most effective preventive measure is drying down to a moisture content in equilibrium with RH 50 or even 35 (Table 8.1), which in some countries may not be difficult. For carry-over stocks, fumigants or contact insecticides may be applied.

The humid tropics present the most difficult situation of all. Provided the temperature and relative humidity remain below 30°C and 70 percent respectively, storage without air conditioning may be possible between harvest and sowing. Over large areas, however, these limits are exceeded and in Malaysia, for example, the relative humidity is usually between 65 and 85 and the temperature between 23 and 31°C, providing ideal conditions for the growth of moulds and insects; natural drying to a moisture content low enough for storage is only possible if the seed is exposed to the sun. Drying in these conditions is limited to small quantities, is slow and erratic, and the germination capacity is liable to be affected. Both fungi and insects have to be guarded against and, in the prevailing temperatures, the seed needs to be dried down to a moisture content in equilibrium with RH 40–50, or to less than RH 40 for carry-over stocks (Table 8.1). Even so, the ambient humidity is so high that the moisture content may rise in store. It is only when the humidity is not too high that cool night air can be drawn through a store to lower the temperature.

The traditional methods of the peasant farmer are adequate for small quantities, but for large-scale storage the costs of the necessary buildings, machinery and energy are extremely high. Any storage beyond the following sowing season involves refrigeration and dehumidification.

Storage of treated seed

The fungicides and insecticides with which seeds are treated involve an element of phytotoxicity which may depress the germination capacity. The phytotoxic effect is exacerbated by cracks and fractures in the seeds, overdosage, inadequate drying and prolonged storage.

Storage of treated seed beyond short holding periods should be avoided as far as possible.

Types of storage

Traditional

The storage methods used by farmers are adapted to the local climate. In the dry tropics and sub-tropics, seed is kept in woven sacks or in heaps on the ground, with protection against animals and perhaps shading against the sun. Protection against rain may not be necessary, but some cover to prevent the condensation of dew and a waterproof underlayer to check absorption from the soil are advisable. In temperate regions, seed is stored in woven sacks or on barn floors, but always under cover. In the wet tropics, more care is taken. Seed is often stored in earthenware bins or in gourds sealed with clay, and these are kept in the house to provide protection against rodents. Baskets are also used, being hung in the kitchen out of reach of rodents and in a dry smoky atmosphere.

Seed may not be kept separately, but simply drawn from the large quantity of grain stored for food. This is usually stored in special huts built of wood, bamboo or straw, with a thatched roof and raised above ground level. Where rainfall is not excessive and the soil impermeable, grain may be stored in pits in the ground.

Buildings

For large quantities of seed, purpose-built stores are necessary. Within the building the seed may be contained in bins, boxes or bags, though bins of the conventional cylindrical type need not be inside a building. If different grades and cultivars are being handled, arrangements must be made to keep them separate. The building should be on a site not liable to flooding and at least 100 metres from any store which might hold grain infested with insects. Fertilizers and chemicals which might injure the seed should not be kept in the same store.

Basically, the building should be of good construction, providing protection against rodents and birds. Floors should be above ground level, and walls should have smooth internal surfaces with no ledges on which seeds may rest. Fans or blowers are necessary for ventilation, and they should have covers which can be tightly closed when not in use. Requirements are more stringent in the tropics than elsewhere. Double roofs and heat insulating materials are helpful; there should be no windows. Ventilation ports should be insect-proof, and the single door should be tight-fitting and kept closed. Concrete, brick and stone protect against rain, but are penetrable by water vapour; moisture vapour proofing may therefore be necessary.

Storage bins are cylindrical in shape to confer strength, and a maximum size of 100 tons is recommended for seed. An aeration system may be incorporated, by which air is blown through a perforated floor and rises through the seed mass. This can be used for an initial drying of the seed after harvest (see chapter 7), or for ameliorating conditions during the storage period. When the ambient air is of relatively low humidity and temperature, aeration can be used to cool the seed or evaporate water which has condensed under the roof. As indicated above, it can even be used to introduce a fumigant.

Seed can be stored in large open-topped wooden boxes, each with a capacity of several tons. These are particularly suitable when numerous small lots have to be handled. The boxes can be placed one above the other to the roof of the store, and can be conveniently stacked, unstacked and carried about the store by fork-lift trucks.

Seed in bags is best stored on pallets, each pallet carrying one ton. Like boxes, pallets can be stacked and moved from place to place by fork-lift trucks. Their use encourages an orderly arrangement with good spacing, adequate aeration and easy access. A stack can easily be enclosed by a sheet for fumigation.

The refrigeration and dehumidification that are necessary for long-term storage in the wet tropics are prohibitively expensive for large quantities but, on a medium scale, insulated rooms fitted with a domestic air-conditioner may be used. These conditioners have the advantage of being universally available and replaceable in case of breakdown. Such stores with a capacity of 100 tons in bags have been successfully operated at 22°C and a relative humidity about 50. It should be remembered that dehumidification creates heat, and the refrigeration unit should be powerful enough to counteract it.

Another possibility for the wet tropics is sealed storage in plastic bags. The densest and thickest grades of polyethylene film can for practical purposes be regarded as impermeable to water, and sealing by heat makes the bags airtight. For this kind of storage, the seed must be drier than for open storage. The seeds respire at a rate dependent upon temperature; this produces water which cannot escape and raises the humidity within the package. Before being put into the bag, therefore, the seeds must be dried down to a low moisture content, in equilibrium with a relative humidity of 30 to 50 depending on the temperature expected, or even lower for long-term storage. During storage the humidity rises to some extent, but not enough to permit the development of moulds. The extra cost of drying can be set against the lesser risk of loss through deterioration.

Routine procedures

Finally, here are six rules to be observed in the management of a seed store. They concern details, but important ones.

1. Keep stocks under continuous observation, looking particularly for hot spots.
2. Fumigate between outgoing and incoming stocks.
3. Keep floors thoroughly swept and burn rubbish.
4. Encourage ventilation within and between stocks. Do not pile seed against a wall.
5. Keep the building in a good state of repair.
6. When seed has to be carried over from one season to the next, select the lots which, in an accelerated ageing test show the best storage potential (see chapter 16).

FURTHER READING

Christensen, C. M. (ed.) (1974), *Storage of Cereal Grains and Their Products*, American Society of Cereal Chemists, St. Paul, Minnesota, 2nd edition.

Harrington, J. F. & Douglas, J. E. (1970), *Seed Storage and Packaging. Applications for India*, National Seeds Corporation, New Delhi.

ISTA (1963), "Drying and storage," *Proceedings of the International Seed Testing Association*, Vol. 28, No. 4.

ISTA (1973), "Seed storage and drying," *Seed Science & Technology*, Vol. 1, No. 3.

Kozlowski, T. T. (ed.) (1972), *Seed Biology*, Vol. 3, chapter 3, New York and London, Academic Press.

MacKay, D. B. & Tomkin, J. H. B. (1967–72), "Investigations in crop longevity," *Journal of the National Institute of Agricultural Botany*, Cambridge. Vol. 11 and 12.

Roberts, E. H. (ed.) (1972), *Viability of Seeds*, London, Chapman & Hall.

9

PROCESSING

SEED AS PRODUCED ON THE FARM MAY NOT BE IN A STATE TO RETAIN ITS viability until the following sowing season. This is mainly a matter of moisture content and is dependent on the weather during and after harvest. Except in regions where good weather is assured at this time of year, provision has to be made for artificial drying to ensure retention of germination capacity.

Furthermore, it is necessary to provide storage conditions in which low moisture content can be maintained, and in which the seed can be protected against the depradations of vermin. Storage requirements depend very much on climate but, even in favourable conditions, storage is a positive need which must be met in organizing a system of seed supply. Important though it be, storage itself does nothing to *improve* the quality of the seed. Apart from providing time for after-ripening, the best it can do is to ensure that the seed at sowing time is as good as it was at harvest time.

However, even if it is dry enough, seed in the condition in which it left the thresher or combine-harvester is not fit for sowing, and a positive improvement in quality has to be made before delivery to the farmer. The most important requirements are cleaning, pesticide treatment and packaging. These operations need to be grouped with drying and storage under closely integrated control, and together they are known as *processing*. Drying and storage are dealt with in chapters 7 and 8; this chapter is concerned with the other operations and with overall management.

For the control of quality, the identity of seed must be maintained all the way through from harvesting to sowing, and this is particularly important during the stages of processing. Processing is not a continuous operation. Seed enters the processing system in lots, passes through in lots, and finally leaves it in lots. It is therefore advisable at this stage to define what is a *seed lot*.

A *lot* is a quantity of seed of one cultivar, of which the origin, history and weight are known. It may be:

(a) Seed harvested from one crop.
(b) A blend of seed harvested from different crops on the same farm.
(c) With the permission of the certification authority, a blend of lots from different farms, each constituent lot being of the same cultivar.

The weight of a lot is usually restricted to 20 tons for cereals and pulses, or 10 tons for smaller seeds, and this is the maximum weight that can be covered by an international certificate.

Seed cleaning

Seed as it leaves the thresher or combine-harvester includes undesirable material which has to be removed. In herbage species it is common for this to amount to 20 percent of the total, but in cereals and pulses it is usually less. The main components of this undesirable material are:

Weed seeds.
Seeds of other crops.
Immature seeds—some of these may be viable, but are poor in germination capacity and have no practical planting value.
Damaged seeds—seeds which are broken, cracked, split, crushed, abraded, or which have been invaded by insects. These also, have a low germination capacity.
Seeds which are too large or too small—these have to be removed to secure uniformity of size.
Other plant material—chaff, pods, awns, pieces of stem, leaves, etc.
Other material—stones, soil particles, rodent faeces, etc.

Some of this material is not positively harmful, but tends to hold moisture and adds to the bulk of material that has to be handled and transported. Moreover, if the ordinary farmer is to be persuaded to pay for high-quality seed, the product he buys must indeed be pure and flow easily through his planter or drill.

To separate these different kinds of material, ingenious machines have been invented which exploit the differences between the physical characteristics of the various components of the mixture. The most important differences are in size, shape and density; separations on these characters are made by perforated screens or cylinders, and by adjustable air blasts. The screens or cylinders are made of sheet metal or of woven wire mesh, and the holes may be rectangular, round, or triangular in shape. Other characteristics that can be utilized are surface texture, colour, electrical conductivity, resilience and affinity for liquids.

Cleaning usually requires a succession of operations and these can be regarded as proceeding in three stages:

(1) conditioning or precleaning
(2) basic cleaning
(3) separation and grading

It may not be necessary for every lot of seed to pass through all three stages. This will depend on the amount of undesirable material present and its nature. If a lot is clean enough after threshing, it may by-pass the first stage and go direct to basic cleaning. A decision on this has to be made when the lot is received at the processing plant.

Conditioning

The purpose of this operation is to facilitate movement of the seed mass through the machines in the subsequent cleaning stages. This is achieved by removing the most bulky material and the rubbish that is most liable to choke up conveyors and sieves. It is a rapid process and consists essentially of an air blast and large-meshed screens or cylinders. The material falls through an upward air stream which removes dust and other light matter. Vibrating screens or revolving cylinders allow particles of the size of the seed to pass through, while chaff, pods, awns, long pieces of stem, leaves, etc., are retained and shaken off to the side. This process of removing large particles is colloquially known as "scalping". In some machines, the process is carried a step further and a proportion of the smaller material is also removed.

Scalping and blowing are operations common to all crops, but certain operations known as "hulling" and "shelling" are necessary for particular crops only. In clovers, the fibrous fruit wall, called a *hull*, adheres closely to the seed and is not always removed by the threshing machine. Removal (or *hulling*) is done by feeding the seed into a cylinder containing revolving arms which rub the seed against the internal concave surface of the cylinder. The machine has to be adjusted carefully to avoid damage to the seeds. This rubbing not only removes the hulls, but to some extent scarifies the hard seeds and renders their seedcoats permeable. Seed analysts remark that the hard seeds which are left at the end of a germination test are smaller than the average. This is because the larger seeds have suffered more severe abrasion during the various cleaning operations and have germinated.

Barley awns and the fibrous tips of oats can be removed by a machine with a similar action, sometimes called an "awner". Removing groundnut seeds from their fibrous pods and detaching maize caryopses from the cob are quite different operations, but both are known as *shelling*. The testa of groundnut is too thin to provide any protection against mechanical damage, and the protruding radicle is especially prone to injury. The testa contains tannin, which is fungistatic, but it is so easily ruptured that this protection is readily lost. This operation is therefore best done by hand, and in some countries the seed is sold to farmers unshelled. In shelling maize, the ears are pressed against revolving cylinders with projecting lugs which, with a thumb-like action, detach the caryopses from the tough fibrous axis. Damage is minimized by delaying this operation until after drying.

After pre-cleaning, the seed may be passed over a magnet to remove pieces of metal.

Basic cleaning

The second stage is a generalized cleaning operation carried out with air blasts and vibrating screens, and is applicable to all kinds of seeds. It is essentially the same as conditioning, but more refined, and carries the process a stage further. It is performed by one machine usually known as an *air-screen cleaner* (figure 9.1). The air blast removes light material. A series of screens separate particles larger and smaller than the crop seed; these may be weed seeds, other crop seeds, or broken seeds. An element of grading can be introduced by including screens which can separate seeds of the crop species which are bigger and smaller than the size required. Screens to separate seeds of different shapes can also be fitted. The success of this operation depends on the selection of screens (over 200 are available) and adjustment of the air blast to suit the species that is being cleaned and the material that has to be removed.

Separation and up-grading

The basic cleaning removes all the adulterants that can be separated by a simple combination of air blast and screens, and for many lots this is sufficient. In some cases, however, adulterants remain which are too close to the pure seed in size and shape to be separated in this way. For such cases other machines are necessary which can separate by other physical characteristics.

In conditioning and basic cleaning, small and shrivelled seeds of the

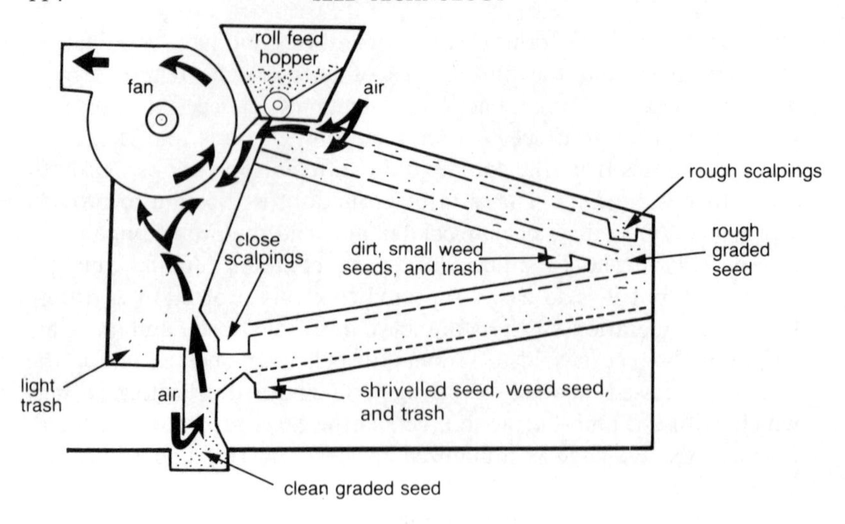

Figure 9.1 Air-screen cleaner (redrawn from *US Department of Agriculture Handbook* No. 179).

species being processed are removed. This may be sufficient for most purposes, but it still leaves quite a wide range in seed size and, for mechanized planting, seed must be uniform. Sizing is the operation of removing seeds larger or smaller than the required size, and special machines may be brought into use for this purpose. Grading is the removal of seeds which are cracked, damaged or otherwise defective. Such seeds may have reduced "vigour" and their presence can influence field establishment.

Seeds of one species are not uniform in their physical characters; they vary, for example, in size and weight. When two species are very close together, their dimensions may overlap; the separation of a mixture will not be perfect, and some of the impurity may remain. Moreover, some of the pure seed is often separated with the impurity, and up to 20 percent of good seed may be lost in this way. The operator must keep the separation process under continuous observation and understand the many fine adjustments that can be made to improve the separation. A seed has three dimensions—length, width and thickness—and to avoid misunderstanding, they can be defined as follows.

Length—the greatest dimension.
Width—the greatest dimension of all possible cross-sections taken at right angles to the length.
Thickness—the greatest dimension of all possible cross-sections taken at right angles to length and to the width.

These dimensions are illustrated for a seed of ryegrass in figure 9.2.

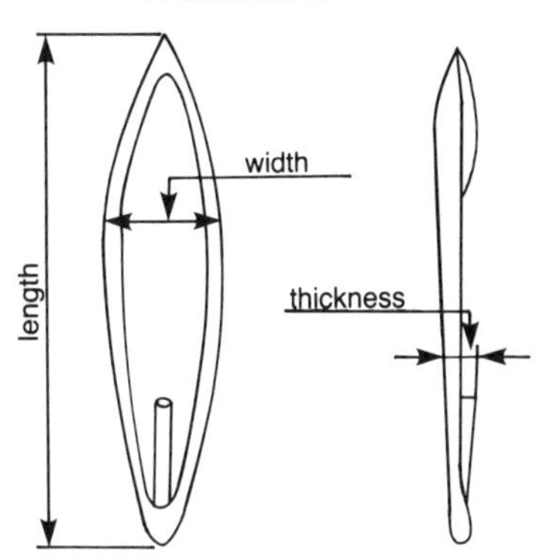

Figure 9.2 The dimensions of a seed, as exemplified by Perennial Ryegrass (redrawn from *Proc. ISTA*, Vol. 34, 1969).

Separation by length

Seeds of similar girth, but of different lengths, e.g. wheat and oats, cannot be separated by a sieve because the longer seed may be tilted from the horizontal position and thus fall through. They can, however, be separated by an indented cylinder. The cylinder has numerous pockets in its inner surface, and revolves round a sloping axle; along its length in the upper half there is a fixed trough. In operation, the seed mass is fed into the higher end of the cylinder and slowly moves towards the lower end. As the cylinder revolves, seeds are carried upwards in the pockets. As the short seeds are carried into the upper half of the cylinder's rotation, they fall out of the indents into the trough, but the long seeds fall out in the lower part of the rotation and so remain in the cylinder. The short seeds are removed by way of the trough, while the long seeds move along the length of the cylinder and emerge from the lower end. The cylinder can also be used for separating seeds of different girth and this operation is illustrated in figure 9.3.

Pulses are liable to be infested by weevils. The larva develops inside the seed and the adult emerges by boring a narrow hole out to the surface. Seeds which have been damaged in this way can be separated by an adaptation of the indented cylinder. Instead of indents, the cylinder has projecting pins which can enter into the holes left by the weevils and so lift the damaged seeds out of the cylinder. As a seed impaled in this way enters into the upper half of the cylinder's revolution, it falls into the trough and is thereby removed.

In another application of this principle, the pockets are cut in the faces of discs which revolve at 30–40 revolutions per minute inside a static cylinder. At the top of the revolution, the short seeds are thrown out into shallow troughs placed parallel to the discs (figure 9.4).

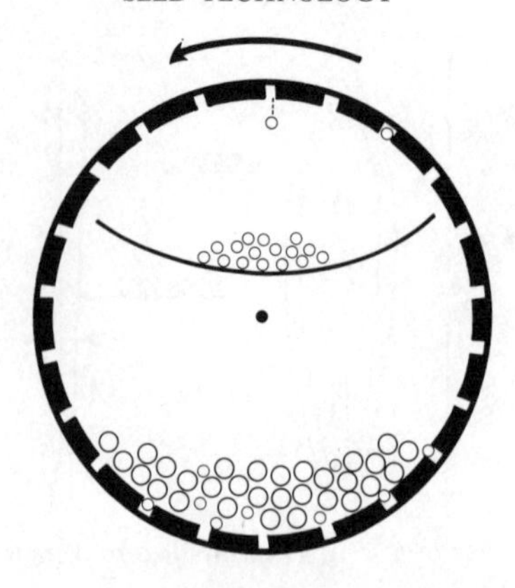

Figure 9.3 Indented cylinder (after Mercer, S. P., *Farm and Garden Seeds*, 1938).

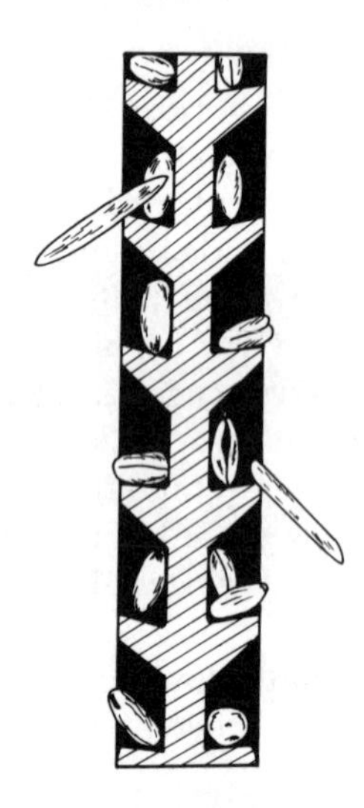

Figure 9.4 Pockets cut in the faces of a Carter disc (redrawn from Carter-Day, Minneapolis).

Separation by shape

One method by which round seeds can be separated from flat seeds is the spiral separator. The seeds are fed on to the top of a spiral chute. Flat seeds slide down the chute relatively slowly and tend to follow the inner edge of the spiral, while round seeds roll down much more quickly, and centrifugal force carries them over to the outer edge, whence they can be led off separately. The seed of purple moon flower, which is shaped like an orange segment, can be removed from the almost spherical seeds of soya in this way.

Separation by width and thickness

This method is used for size grading rather than for separating seeds of different species. Holes of the required cross-section are set in screens or cylinders, e.g. rectangular for maize and round for barley. A seed cannot pass through if it is lying horizontally; therefore each seed has to be stood on end or on edge in order to present its appropriate cross-section to the hole. To achieve this, the holes are set in grooves for flat maize or in round pits for barley (figures 9.5 and 9.6).

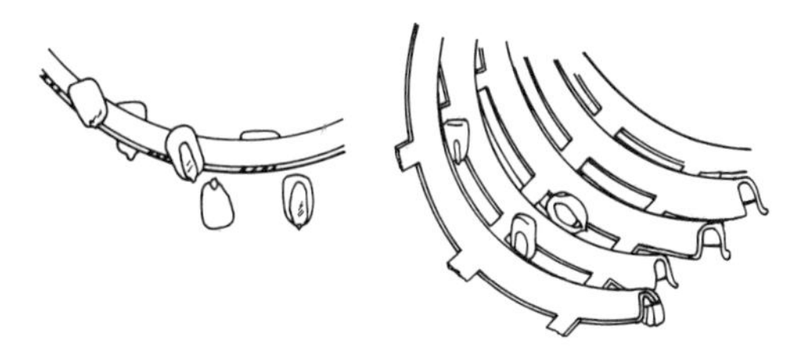

Figure 9.5 Cylinder with rectangular slots in grooves for separating maize kernels by thickness. The saddle between the grooves up-edges the kernel and presents it to the slot in the appropriate position (redrawn from Carter-Day, Minneapolis).

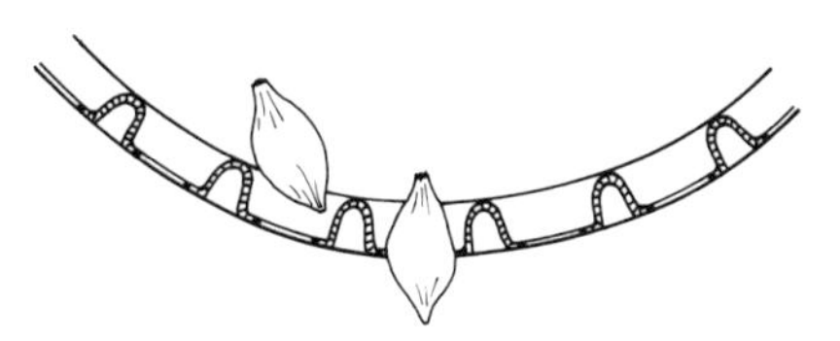

Figure 9.6 Cylinder with round recessed holes for separating barley grains by width. The grain is up-ended in the recess and presented to the hole in the appropriate position (redrawn from Carter-Day, Minneapolis).

Separation by specific gravity

Seeds of similar size and shape, but slightly different in weight, can be separated by the *gravity separator*. This is used for separating empty from full seeds of flax and various pulses, for example. The seed is spread in a layer over a perforated plate (or deck) set at a slope. A current of air is blown vertically through the deck, sufficiently strong to lift the light seed slightly off the surface while the heavy seed remains firmly on the deck. The deck shakes backward and forwards in the direction of the slope with sudden jerky movements. Each jerk moves the heavy seed slightly up the slope. The light seed, floating on air, is unaffected by the movement, and tends to move down the slope under the influence of gravity. Eventually there is a separation, with heavy seeds at the top of the slope and light seeds at the bottom.

Another machine which can separate seeds differing in specific gravity is the oscillating table separator. This consists of a table with zig-zag partitions along its length (see plan in figure 9.7). The table is

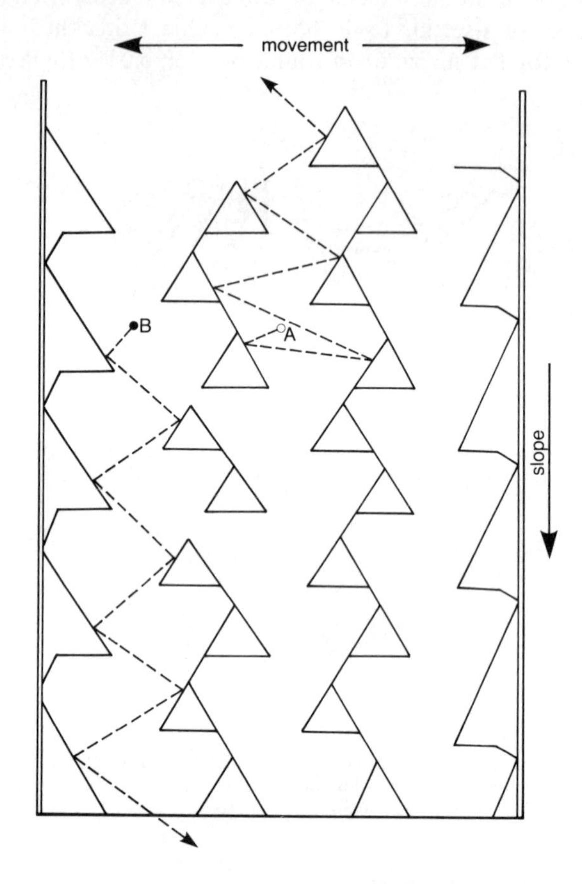

Figure 9.7 Oscillating table separator (after Mercer, S. P., *Farm and Garden Seeds*, 1938).

inclined longitudinally and is shaken with a jerky movement from side to side. The seed flows slowly on to the table and tends to slide downwards, but is rhythmically struck against the partitions by the lateral movements of the table. In addition to lateral forces, the seeds are subjected to two forces acting up and down the slope—the impact of the diagonal faces of the partitions acting upwards and gravity acting downwards. In the case of a light seed (A in figure 9.7), the impact of the partition is sufficient to throw it across to the opposite diagonal, which gives it another thrust upwards. The seed therefore tends to move upwards. In the case of a heavy seed (B), gravity is sufficient after impact to take the seed down the slope, so that it misses the opposite diagonal and strikes the next lower one. The seed therefore tends to move down the slope. A complete separation is possible, with light seed moving off the upper end of the table and heavy seed moving off the lower end. To obtain a good separation, the slope of the table and the speed of the oscillation are adjusted according to the size of the seeds and the difference between them.

These two machines can be used to improve the quality of a seed lot by removing seeds of the same species but of low specific gravity. Such seeds are poorly developed and have low planting value.

Separation by surface texture

Seeds with a rough seedcoat adhere to velvet cloth and can be separated from smooth seeds in this way. This method is commonly used to remove rough weed seeds, such as dodder and cranesbill, from the smooth seed of clovers and lucerne. The seed is poured on to a piece of cloth moving upwards, on a roll or otherwise; the smooth seeds slide down the slope and are collected at the bottom, while the weed seeds are carried upwards (figure 9.8). The same principle can be applied to larger seeds, and even to the removal of wild oats from cereals.

This idea is carried further by the *friction separator* in which the seeds and impurities are brought into contact with two surfaces, one rough and one smooth. This machine was designed primarily to remove stones and clods of soil from beans, but can also separate immature and damaged seeds.

Separation by colour

Separation by colour is used not so much for separation of species as for the removal from pulses of discoloured seeds which might be diseased. In the machine, the seeds fall singly in front of a group of photo-electric cells which are activated by any variant from the normal colour. This triggers off a mechanism which diverts the variant seed from the falling stream, so that it leaves the machine through a different channel. In one model, the seed is diverted by a brief blast of air, directed so as not to affect adjacent seeds in the stream (figure 9.9). In another type, the variant seed is given an electric charge; the seed then falls through an electric field which diverts the charged seeds.

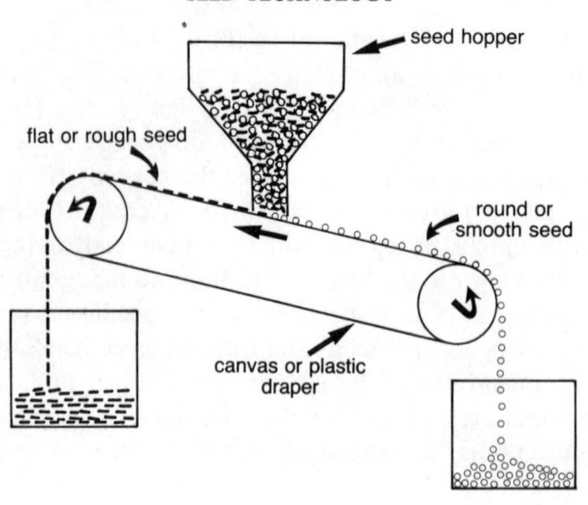

Figure 9.8 Inclined draper separator (redrawn from *US Department of Agriculture Handbook* No. 179).

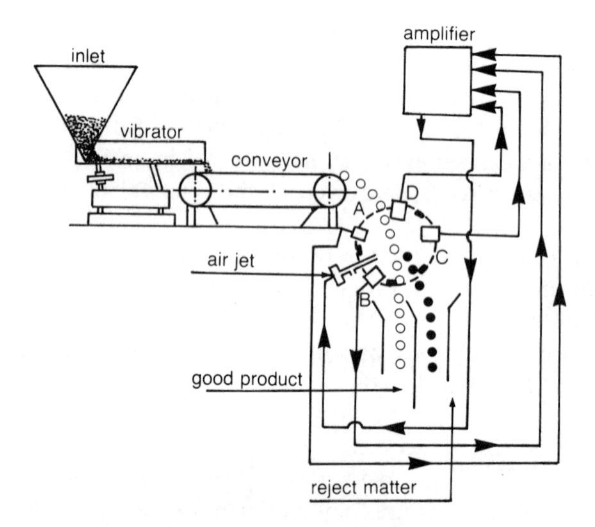

Figure 9.9 Diagram of an electronic colour separator. A, B, C, D—photo-cells (redrawn from *Proc. ISTA*, Vol. 34, 1969).

Separation by electrical conductivity

When seeds are given an electric charge, they may lose it quickly or retain it for a time, depending on their conductivity. To exploit this difference, the seeds are passed through an electric field immediately after being charged; the charged seeds are diverted, while seeds which have lost their charge fall straight through. Alternatively, the seeds may fall on to an earthed conveyor belt, to which only the charged ones adhere. Docks can be separated from clovers in this way.

Separation by affinity for liquids

Some seeds are more easily wetted than others, and this difference can be used in separation. This character is, however, associated with surface texture, and separation on a rough cloth (as for dodder) may be more convenient. In this method the seeds are sprinkled with water and then dusted with iron powder or with a light powder such as sawdust. The powder adheres to the wet seeds and they can be separated either by a magnet or by means of the gravity separator. Iron powder will stick to cracks and damaged areas of seedcoat; clover seed, for example, can be upgraded by this method.

Separation by resilience

Even differences in resilience, or the ability to bounce, can be used. The seeds are dropped on to a smooth sloping surface; seeds which do not bounce, slide off the edge, while seeds which can bounce are thrown a short distance beyond the edge and fall into another chute. This method can be used, for example, to separate clover seeds from seeds of *Poa*.

Blending

Ideally, if samples are taken from different bags within a lot, or from different positions within a bag, the purity, weed content, germination capacity, etc., should be the same in all samples. In practice, however, there is always some variation and such a state of perfection is never attained. Some variations within a lot is therefore acceptable, but investigations have shown that it may be excessive in about 10 percent of lots sold to farmers.

There are two causes of variability within a seed lot. The first is variability within the crop from which the seed was harvested. There is never complete uniformity throughout a crop, and the seed harvested from different parts of the field varies in maturity, germination capacity, weed content, etc. Seed from different parts of the field tends to become mixed as it passes through the various processing machines, but even at the end of the line there is some residual variation. This may be rectified by putting the seed through a blending process, but in practice this is seldom done for a lot derived from one crop.

The second reason is that lots from different fields or from different farms are blended together, but the blending process is ineffective and fails to mix the component parts thoroughly. The seed lots of any particular cultivar arriving at a processing plant from the farms differ greatly in size. Management is simplified and the plant works more efficiently if they can be combined into larger bulks and packaged and sold in lots of uniform size. Sometimes a lot of poor quality is blended with one or more high-quality lots of the same cultivar in proportions calculated so that the average quality of the mixture is up to the required standard. The ethics of this procedure are arguable, but it can be justified only if the blending is thorough enough to ensure that every bag in the lot meets the standard.

The aim in blending is to mix the constituents so thoroughly that when the composite lot is packaged, they are present in each bag in the proportions in which they were introduced into the mixture. The extent to which this is achieved in practice depends on the diversity between the constituents, the quantity of seed, and the method employed.

The greater the difference between the constituents, the less chance is there of uniformity being achieved. The lots which are blended together should therefore be as like each other as possible, and this has to be borne in mind in mixing lots of different quality. A small lot is more likely to be uniformly mixed than a large one, because the blending requires less time and mechanical effort, and because the mixing chambers and bins available in processing plants are limited in capacity. The international rules for seed testing prescribe a maximum size of seed lot, the limit being 10 tons for seeds smaller than wheat and 20 tons for larger seeds. This limit is, of course, arbitrary but it recognizes that in a large lot there is liable to be so much variation between bags that the quality of the lot cannot be represented within acceptable limits by a single sample.

Various machines and procedures are used for blending, but uniformity is not always attained. Our understanding of the operation is inadequate. Mixing involves movement of particles induced mechanically or by gravity; but there is not sufficient theoretical or experimental knowledge of how the effectiveness of this movement is influenced by direction and speed, or how it is modified by the physical properties of the seeds. The usual blending procedure for large lots of one species is to introduce the components simultaneously into a bin, and then circulate the bulk two or three times between bins. For smaller lots, the seed is run into a mixing chamber and stirred around by a revolving screw or by rotation of the whole chamber. The screw is less effective than movement of the chamber and is liable to damage the seeds.

A blending bin designed on the basis of experimental work in the

United States promises to be more effective. The idea is to fill the bin with the seed to be blended and draw seed from different levels at the same time, so that seed from each level flows into another bin along with seed from every other level. The bin is four-sided, with an external duct running from top to bottom along the centre of each side (figure 9.10). In each duct, at equal intervals along its length, are

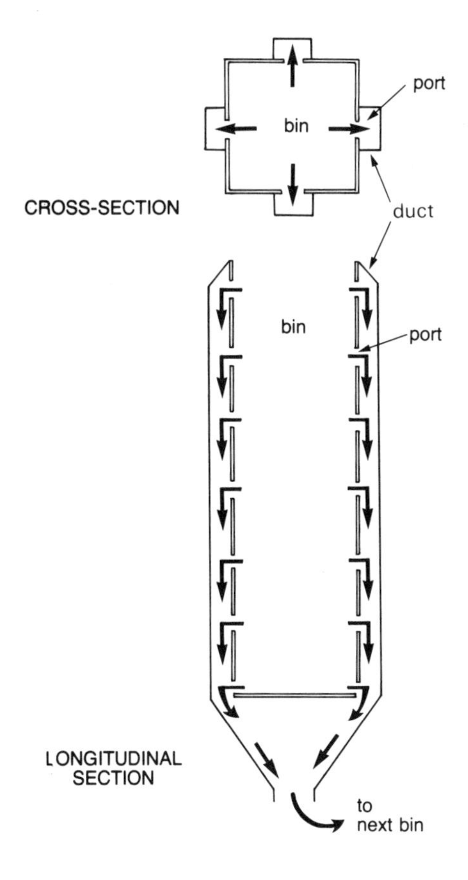

Figure 9.10 Blending bin.

circular ports through which seed can flow from the bin into the duct. The bin is filled with the ports closed. All the ports are then opened simultaneously, and the seed runs down into a hopper at the base of the bin. From the hopper, the seed is carried to a second similar bin and the operation is repeated. After passing through three blending bins, the composite lot is sufficiently uniform.

A farmer may require a mixture of seeds of different species, or of different cultivars, to be delivered to his farm ready to sow. In the former case, because the physical properties of the constituents differ, they are liable to segregate again after blending has been completed. The contents of each bag are therefore mixed independently in a small mixing chamber. The constituents are held in large storage bins and measured electronically in the proportions required into the mixing chamber. After mixing, the seed is transferred automatically to a bag.

Treatment

Pesticides are applies to seed as means of controlling pests and diseases in the crop. They have three modes of action:

to destroy pathogenic organisms on (or just within) the surface of the seed, e.g. covered smut, or bunt, of wheat.

to protect the emerging seedling against insects and fungi in the soil at sowing time, e.g. flea-beetles and damping-off fungi.

to be absorbed by the seedling at sowing time and thus destroy organisms that were within the embryo (e.g. loose smut of barley) or protect the plant against subsequent air-borne infection by foliar diseases such as mildew.

Fungicides are ineffective in preventing the growth of mould fungi, which occurs when seed of high moisture content is stored. Nevertheless, they can check to some extent fungal growth in seeds which are particularly prone to mechanical injury. For example, groundnut seeds are very liable to be injured in the shelling process, and the wounds and abrasions provide entry points for storage fungi, causing discoloration and decay; storage life can be prolonged by treating with a fungicide immediately after shelling.

Of the commonly used fungicides, the organo-mercurial compounds are most effective against infection carried on the seed surface. These compounds are applied in powder form, but they are volatile, diffusing throughout the seed mass; to allow this to happen, seed should be left in closed containers for 24 to 48 hours after treatment. They are commonly used on small-grain cereals, but are phytotoxic and overdosed seeds are liable to produce abnormal seedlings. They have a bactericidal effect and should not be applied to seeds treated with a bacterial inoculant.

Non-mercurial compounds, such as thiram and captan, are non-volatile and for this reason are not effective against surface spores, but provide protection against fungi in the soil. They are not phytotoxic and are commonly applied to the seed of beans, groundnuts, vegetables and maize.

Pesticides are potentially dangerous and have to be used with care. Indeed, in some countries the use of certain compounds is illegal. They may be extremely poisonous to human beings and animals. Workers in the processing establishment are at risk, and can be affected if the material comes into contact with the skin or is inhaled. Some of the substances used are volatile, but even non-volatile materials may be dispersed in the air. The application of pesticides to seed is done in closed machines, and as a general precaution they are stored in closed containers. Extractor ventilating fans operate when seed is being treated; workers wear protective clothing, including masks, and wash thoroughly after every period of continuous operation—particularly before eating, drinking or smoking.

The farmer who opens the bags and sows the seed is also at risk, but only briefly as compared with the process operator. Nevertheless, he needs to be warned; treated seed should be distinctively coloured and marked as dangerous—not only in words, but in vivid symbols. He should be advised particularly not to open bags, except at the actual time and place of sowing, not to store the seed near human or animal food, and on no account to use surplus seed for food. Disposal of surplus seed at the end of the season can be a problem for a seed company, but it must certainly not be used for human or animal food.

The active ingredients of pesticides are, of course, poisonous to fungi and insects, but in certain circumstances they can also be poisonous to the seeds. The toxic effects can be seen at germination in the abnormal growth of the seedlings. Typical symptoms in cereals are short thickened roots and plumules (figure 16.6). Water absorption is restricted and, if the soil is partially dry, the seedling dies from drought; but if the soil is wet enough, sufficient water may be absorbed for the seedlings to recover and grow normally, though somewhat delayed.

The problem lies in applying enough pesticide to the seed to be fungicidally or insecticidally effective without impairing germination. In some cases the effective dosage is substantially below the toxic dose, but in others it is very close and there is little margin for error. A precise dose has therefore to be applied uniformly over all the seeds in the lot.

The harmful effects on the seed depend not only on dosage, but also on certain other factors—moisture content, state of the seed coats, and storage. When the dosage is right, these factors, if adverse, may nevertheless induce harmful effects, while the consequences of over-dosage are exaggerated.

The maximum safe moisture content for cereal seed that is to be treated is 16 percent in a temperate climate, but it can be assumed that, if the seed is dry enough for between-season storage in local conditions, it can safely be treated. If the testa or other seed covering is broken, an excessive amount of the pesticide is liable to enter the seed and be absorbed by the embryo.

Harmful effects may develop if the seeds are stored for too long a period or at too high a temperature. As these factors intensify rather than cause the trouble, it is not possible to prescribe precise limits. As a general rule, seed which is expected to be stored beyond the following season should not be treated with pesticides of this kind; they may nevertheless, with due precautions, be treated with contact insecticides to control storage pests. In practice, little can be done to control the storage temperature of normal seed supplies, and in tropical countries high temperature may rule out the possibility of treatment before storage, or restrict the choice of pesticides. In these circumstances it may be advisable to issue a relatively harmless formulation in small quantities to the farmers and let them apply it themselves at sowing time. This, however, is a matter of local policy to be decided after consultation with the experts.

Pesticides are produced in powder or liquid form. Powder may be applied dry or, to eliminate flying dust, mixed with water to form a slurry; liquids as a spray or mist. For safe and efficient application, the process needs to be done mechanically; the operators must not handle, and need not even see, the seed. Precise instructions are issued by the manufacturers of the machines and of the pesticides, and these must be strictly followed. In some treatments a dye is introduced to give the seed a distinctive colour. The treatment may be applied to a continuous flow of seed or in batches. In the continuous method, the pesticide is measured on to the seed as it is moved through an inclined cylindrical mixing chamber by an auger. In the batch process, enough seed to fill a bag is introduced into a mixing chamber along with the appropriate dose of pesticide and mixed by a rotor; the entire contents of the mixing chamber are then discharged into a bag. Tests have shown that the application of powders by the continuous process is sometimes inefficient, only about one quarter of the total application still adhering to the seeds when they are planted in the soil. The batch method is to be preferred.

Another treatment that can be applied to seeds is pelleting (chapter 1). The material used as coating must be porous in some degree to permit respiration by the seed in store, but should not absorb moisture from the atmosphere. In the soil, however, it must absorb water and soften rapidly so as not to impede swelling of the seed and emergence of the seedling. Fertilizers and pesticides can be incorporated in the coating material to promote establishment and, for clover seeds, a culture of the *Rhizobium* organism may be added, provided sowing is done soon afterwards. Materials sometimes used for pelletting are finely powdered bauxite, clay, lime or rock phosphate with gum arabic or resin as an adhesive.

Seeds may also be enclosed in granulates or tapes. These treatments are expensive, but some farmers and growers are prepared to pay the cost, which is offset by lower labour costs in the field, by the smaller number of seeds that has to be purchased, and by convenience.

Packaging

At the end of the processing line the seed is packed into containers of uniform size, which are then closed. In supplying seed to farmers, the form of container universally used is the bag (figure 9.11).

The transfer of the cleaned seed from the processing plant to the field where it is to be sown is neither a simple nor a speedy operation. The seed may have to travel long distances by a variety of means—rail, lorry, animal-drawn carts, on animals' backs or even on men's backs—and be subjected to jolting and rough handling. Its journey may be interrupted by periods of storage in country merchants' or village traders' premises. On arrival at the farm it will be kept for a time in a barn, in the farmer's house or possibly out-of-doors. Throughout all this, there must be no leakage of the pesticide with which the seed has been treated. The package should therefore be regarded as:

a convenient unit for handling, transport and storing
a protection against contamination, mechanical damage and loss
a suitable environment for storage
a barrier against loss of seed and the escape of pesticides
a sales promoter

Bags may be manufactured from cloth, paper or plastic film.

Cloth bags are woven from natural materials such as jute or cotton, from synthetic fibres such as nylon, or from a mixture of different fibres. Because of their texture, they can be built into stable sacks and are easily carried on men's backs. They provide good protection against rough handling, but are easily torn by rodents. Being permeable, they allow the seeds to absorb or lose moisture according to the humidity of the ambient atmosphere; they are therefore suitable for storage only in dry or cool conditions. A minor advantage is that samples for test can be drawn with a spear without damaging the material. They are not suitable for seed treated with a highly poisonous pesticide.

Paper bags are built up of several thick layers and are surprisingly strong. Layers of plastic material are sometimes incorporated to make them more suitable for humid conditions. If a sample is drawn through a spear, the hole can be repaired with an adhesive patch.

Plastic film, especially polyethylene, is being increasingly used for

Figure 9.11 Dispatch of packaged seeds from warehouse to farms (Scottish Agricultural Industries).

packaging. Its advantages are strength and impermeability; both these characters vary with the density and thickness of the film. Its disadvantages are its smooth surface, which allows bags to slide over each other so that stacks collapse, and the difficulty of drawing samples. The denser and thicker grades provide good protection against rough handling and are resistant to rodent damage.

This material would seem to be ideal for providing sealed-storage conditions in a humid climate, giving protection against high humidity in a seed store, during transport, in merchants' premises and on the farm, and even against rain when left out-of-doors. Packaging in sealed bags, however, requires additional drying, and the cost of this cannot be justified in a temperate climate. To overcome this, small holes can be made in the bags to permit diffusion of water vapour, but this in turn incurs the disadvantage that the bags cannot be left exposed to rain.

Bags may be closed by hand-tying with string, but in large-scale operations a stitching machine is used. Polyethylene bags must be heat-sealed if sealed storage is required. Paper and polyethylene bags can be closed by a valve, which is shut automatically by the weight of seed when the bag is turned over after filling. For certified seed, an official sealing device is incorporated with whatever closing method is used.

Some information about the contents of the bag must be displayed. This should include at least the species, the cultivar, the grade and a lot reference number, but the law may require more details. This information may be printed on the bag itself, or on a label or tag firmly attached as part of the stitching or sealing operation.

The bag should also be regarded as a means of promoting sales, and this minimum amount of information can be supplemented by a bold distinctive design and, for high-technology cultivars, perhaps some advice on cultural methods. If the seed has been chemically treated, this should be indicated by some dramatic emblem.

The size of the package should be adapted to local conditions. After leaving the processing establishment, loading, unloading and movement over short distances are likely to be done manually, and this sets the upper limit as the weight that a man can carry. The capacity may also be adjusted to provide enough seed to sow a standard area—in local terms, e.g. 22 lb of maize for one acre. However, packaging is costly, depending on the size of the unit, and may become prohibitive if the packages are too small.

Process management

The normal sequence of the operations described in this chapter are as follows:

RECEIVING—PRE-CLEANING—CLEANING—SEPARATION—BLENDING—TREATMENT—PACKAGING—DISPATCH

In the progression of a seed lot through a processing plant, two other important operations have to be integrated into the sequence—drying and storage—and these are discussed in other chapters. At what point in the line these operations are inserted depends on the kind of seed, the condition of the seed lot, and the work load.

A seed lot remains within the processing plant for weeks or months, but only a fraction of this time is spent in actually moving through the machines. For most of the time it is stored and, provided it is dry enough, this may be at any point in the sequence. It can be stored in bulk or in bags as received, in bins or large boxes between operations, or in bags after packaging. It may be put into store because the next stage in the sequence is full to capacity, or moved out of store because space is required for another lot. It is an operational problem.

Drying may also be done at different stages in the sequence, and need not be completed in one operation. Seed may be partially dried on the farm but, if it is received with a moisture content more than 3 percent above the level considered safe for storage, should be dried immediately. Seed which is close to the safe moisture content may be cleaned wholly or partially before a second drying, but at least it should be pre-cleaned to remove the coarse trash. An advantage of this procedure is that the seed is less susceptible to mechanical damage before it is completely dried. Seed destined for sealed polyethylene bags should be given a final drying as short a time as possible before packaging.

The machines used for the various operations are installed in sequence in a processing line, but with arrangements for by-passing any machine if that particular operation is not required. The capacity of all operational stages should be similar, so that the seed can move steadily along the line without delays. For a smooth flow, a conveyor system is necessary.

The machinery may be installed in a multi-storey building, with a different stage on each floor. The incoming seed is elevated to bins on the top floor, flows downwards by gravity from stage to stage, and on the ground floor is finally packaged in readiness for storage and dispatch. A horizontal arrangement with everything at ground level and a separate elevator for each machine is, however, less costly to construct and easier to supervise in operation.

In planning the layout and setting up machines, it should be remembered that all seeds are to some degree fragile, though some, particularly pulses, more than others. The consequences of breakage are obvious, but cracks and abrasions in the seedcoat, though not apparent to the eye, may have indirect effects by providing access for micro-organisms and toxic chemicals. In an experiment in the United States, the germination capacity of snap beans declined from 95 to 53 percent when they were dropped on to a hard floor from a height of about two metres; after nine such drops, the germination had fallen to zero. Mechanical damage is most frequently caused in the threshing and shelling processes, due to faulty adjustment of the machines or unsuitable moisture content. Thereafter, damage is commonly caused during processing by conveyors and drops.

A seed-producing establishment is a busy place with complex movements and operations continuously in progress. Seeds of different species and cultivars are received, different lots are subjected to different operations, weeds and rubbish have to be extracted and disposed of, spillage is inevitable, and all this may go on non-stop day and night, with shift changes among the workers. In such conditions, chaos is possible—good seed becomes contaminated, cultivars are mixed, while an accumulation of spillage and trash provides perfect conditions for insect and rodent infestation, and presents a continual fire risk. All this has to be kept in mind in designing the building, in planning the layout, and in daily management.

The building should be of solid construction and spacious. The machines must, of course, be placed in relation to each other and not too far apart, but adequate space is required between them for operations such as changing screens and cleaning, and for the movement of trolleys and fork-lift trucks. Space for stacks of boxes or bins should be sufficient for the stacks to be widely separated. Floors should be smooth to facilitate cleaning. Walls should be free of crevices which might harbour insects, and of ledges on which dust and fragments of plant material can accumulate. Bins and boxes should have smooth surfaces with no cracks or other irregularities in which seeds might lodge and possibly contaminate subsequent lots.

Good management is characterized by orderliness and cleanliness.

The work load of a processing establishment is not uniform throughout the year. Active periods start at harvest time and end when the packaged seed is dispatched, and there may be more than one such period in a year, distinct or overlapping. For each season a plan of operations should be prepared in advance, but the plan must be flexible enough to allow for contingencies such as an unusually early harvest, or unexpectedly high moisture contents in incoming seed lots.

Laboratory tests are carried out on samples drawn before and after processing, and at intermediate stages, and these are supplemented by the observations of experienced operators. The most important test is the initial one to determine moisture content and identify the impurities that have to be removed. On this basis a programme of operations can be prepared for each lot. Throughout the processing, the separate identity of the lot is maintained with clear labelling and wide separation. It is helpful if in all operations, species and cultivars can be kept separate in both time and space.

Records are important. As each lot is received, it is given a reference number, and a record is kept of its previous history, of every process it goes through, and of its destination when it leaves the premises. This procedure makes it possible to disprove or confirm reports that are received of faults in the product, and to investigate the cause. If, for example, a farmer complains that seed which he has purchased has failed to germinate in the field, it is possible through the records to trace other farms where seed from the same lot has been sown. If the failure is confined to one farm, it can be concluded that it is probably due to some fault or accident after delivery. If, however, all sowings of seed from the same lot have failed, the fault must lie in the seed, and records may enable the cause to be discovered. Apart from this value to the seed company, detailed records may have to be kept in order to comply with the seed laws or with the regulations of the certification authority.

Each machine is thoroughly cleaned on changing the species or cultivar being processed, or on changing to a higher grade of the same cultivar. In processing cotton for seed, the use of each ginning machine is restricted to one cultivar. The precautions and procedures necessary for harvesting machinery have been described in chapter 7, and the same principles apply in a processing establishment.

Infestation by rodents and insects has to be guarded against. These pests thrive in sheltered spaces left undisturbed for long periods, and for this reason stacks are separated by clear spaces from each other and from walls. At least once a year the premises should be sprayed with a contact insecticide, and an intensive rodent extermination exercise carried out.

A processing establishment produces large quantities of waste material, consisting of impurities removed from the seed by the machines and spillage on to the floor. It includes seeds, and there is a risk of contamination unless a regular system of disposal is followed. Floors should be swept every day and, less frequently, horizontal surfaces on beams and ledges where seeds may lodge, should be cleaned. For all cleaning operations, compressed air and powerful vacuum cleaners are invaluable. Dust extractors continuously operating can remove fine particles floating in the air. To destroy weed seeds in the waste material, it can be treated with a hammer-mill and used for livestock feed or compost; alternatively, it may be burned. As discarded seeds may be infested with insects which can transfer to sound seeds in store, the disposal should be completed quickly. There may be legal restrictions on the disposal of waste but, whatever method is used, pollution in the vicinity of the building should be avoided.

With so much inflammable material on the premises, there is an ever-present risk of fire. This can be minimized by efficient waste disposal and dust extraction, and the high value of the material justifies covering the risk with an insurance policy.

The efficiency of the machinery is very dependent on the skill of the operators. It is not sufficient to leave apprentices to acquire this skill by working under a more experienced man; they need special training courses and periodic checking. The environment inside a processing plant is highly artificial, and all workers need protection in some degree—from injury by moving machinery, from breathing dust-laden air, and from poisoning by toxic chemicals. An enlightened management enforces safety and health precautions voluntarily, but in some countries these are enforceable by law.

As indicated above, processing machinery at certain seasons is working all day and every day. At such times a breakdown can have serious consequences, bringing the whole establishment to a halt—briefly if repairs can be executed immediately, or for a prolonged period if spare parts have to be brought from a distance. The likelihood of a breakdown can be minimized by following the manufacturer's service specifications. These will include systematic lubrication, regular inspection, periodic replacement of parts particularly liable to wear, and a complete overhaul annually during a quiet period. Spare parts have to be kept available, and the stock maintained will depend on an estimate of probable needs. A large establishment may employ a staff of competent engineers capable of carrying out a servicing programme and of dealing with emergencies. For smaller concerns, however, it is better to arrange a servicing contract with local engineers or the manufacturer's agent, and to employ mechanics for lubrication and minor adjustments only.

Seed processing is subject to commercial competition and is such a complex business that it cannot be left to run itself; active management, alert to any possibility of improving efficiency, is necessary. Inputs and outputs need to be kept under close scrutiny for the purpose of monitoring efficiency, and each year's performance should be analysed and compared with those of previous years.

FURTHER READING

FAO (1974), "The Use of Mercury and Alternative Compounds as Seed Dressings." Report of a joint FAO/WHO meeting.

Gregg, B. R. and others (1970), *Seed Processing*, National Seed Corporation, New Delhi.

ISTA (1969), "Equipment Number," *Proceedings of the International Seed Testing Association*, Vol. 34, No. 1.

"Seed Cleaning and Processing Number," *Seed Science & Technology*, Vol. 5, No. 2, 1977.

Vaughan, C. E. and others (1968), *Seed Processing and Handling*, Handbook No. 1, Seed Technology Laboratory. Mississippi State University.

Vaughan, C. E. and others (1972), "Further studies on blending free flowing seeds with bin flow control devices," *Proceedings International Seed Testing Association*, Vol. 37, p. 681.

Virdi, S. S. and others (1972), *Seed Processing Plant Operator's Handbook*, National Seed Corporation, New Delhi.

10

MULTIPLICATION

Cultivars

The concept of a cultivar is not so precise as that of a species. There are exceptions, but in general one species does not cross-pollinate with another; intermediate forms are not produced and there is no doubt to which species a crop plant belongs. Within a species, however, there may be many cultivars; they are usually interferile and crossing can produce intermediate types. The capacity of a cultivar to maintain its distinctive features depends very much on its mode of propagation and its breeding system.

Vegetative propagation

In some crop species, propagation is vegetative, e.g. potato, banana and yams. In these crops a piece of stem taken from the mother plant is planted and develops into an independent unit. In one cultivar of this type, every plant, though growing separately, has been derived vegetatively from one original plant—as though all the branches of a large tree were growing separately with their own root systems, instead of being connected to the main trunk. Such a cultivar is called a *clone* and all plants belonging to the same clone are identical. Mutations may occur, though rarely, but otherwise, so long as the cultivar is propagated vegetatively, it remains unchanged. This book, however, is not concerned with vegetatively propagated crops, but only with crops which are propagated by seeds.

Breeding systems

A seed develops in the ovary of a flower after a sexual fusion. The two elements in this fusion may come from the same plant (self-pollination), or from two different plants (cross-pollination).

In a self-pollinated species, such as barley and wheat, all the plants in a cultivar are descended from one common ancestor by a process of repeated self-fertilization and are therefore genetically homozygous. Every plant is identical with every other plant within the cultivar and produces identical progeny in the next generation. Such a cultivar is called a *pure line*; provided cross-pollination does not occur, it can maintain its distinctive characters over an indefinite number of generations. Nevertheless, from time to time irregularities may occur in the pairing and separation of chromosomes, and give rise to atypical plants.

At the other extreme are cultivars in which seed is normally produced by cross-pollination. The individual plants may be completely self-sterile or, if self-fertilization happens, few normal seeds are produced. In such a cultivar, the individual plants are genetically heterozygous and are not identical with each other. The cultivar is a mixture of plants differing in inherited characters such as habit of growth and maturity. In some cultivars the range of variation is very narrow, but in others it is quite wide. Because the plants intercross with each other, an individual is not the same as either of its parents, but the progeny from a group of plants shows a range of characters roughly similar to that of all the parents. Provided pollen from another cultivar is not introduced, the cultivar as a whole can remain broadly unchanged over several generations.

Cultivar purity

It follows that the concept of purity cannot be the same for both self-pollinated and cross-pollinated cultivars. For a pure line, a precise description can be prepared, and any plant which does not fit the description can be regarded as an impurity. For a cross-pollinated cultivar, the description is more general, allowing for variation between individual plants, and there is no well-defined boundary line beyond which any plant can be called an impurity. An individual plant can be regarded as an impurity only if it is clearly and significantly outside the accepted range of variation. In practice, decisions on cultivar authenticity are made on an estimate of the average value of measurable characters and the statistical variation.

The contrast between these two kinds of cultivar is well illustrated by the seed certification standards of the European Economic Community. In cereals, a crop grown to produce basic seed is required to have a cultivar purity of 99.9 percent, calculated by number of plants. For forage crops, however, there is no precise purity standard but (apart from gross off-types) the crop is required to be "satisfactory for trueness to cultivar".

In multiplying seed, the purity, or trueness to cultivar, of each generation has to be confirmed. This is done by examination of the seed in a laboratory, or of plants growing in the field or in a check plot. The essential features that distinguish cultivars are characters of interest to the farmer or to the ultimate user, such as yield, straw strength, cooking quality and palatability. For a number of reasons, such characters cannot be used in laboratory tests, plot examinations, or field inspections—the work involved may be excessive or the tests too prolonged, the characters may not be manifest at the required time, or may be influenced by weather conditions. For control of cultivar purity, the cultivar should have characteristics by which it can be clearly and simply distinguished. In some cases agronomic characters, such as habit of growth and flowering time, can be used. In other cases morphological characters are used, such as the rachilla hairs in barley and the shape of the glumes in wheat. These characters are of no significance to the farmer or the ultimate user, but are associated with the more essential features; they are important because they serve as markers and facilitate the control of seed multiplication.

The most useful markers are "two-state" characters. Such a character has two alternative conditions, e.g. coloured or colourless, awned or awnless, and no other possibility exists. Two-state characters are more common in self-pollinated cultivars. Other characters are "multi-state". Such a character varies over a range, with intermediate states lying between two extremes, e.g. stems short to long, growth habit prostrate to erect, date of flowering spread over several days. These multi-state characters are typical of cross-pollinated cultivars; the variation within a cultivar is not so much due to the presence or absence of different characters as in the range of expression of each character.

It must not be supposed that every species is either self- or cross-pollinated. In many species both occur, though one or the other may predominate. In such species the amount of variation within a cultivar depends on the degree of self-fertility. If a species is highly self-fertile, a flower sets seed whether the pollen comes from the same or from a neighbouring plant and, by a process of selection, a cultivar can be produced which is as uniform, or nearly so, as a pure line. If, however, self-fertility is low, the majority of seeds are the result of cross-pollination, and the same degree of uniformity can never be attained. As a general rule, uniformity within a cultivar is associated with self-fertility.

F1 hybrid cultivars have been developed, particularly in maize. On the maize plant the flowers are unisexual, and the male and female flowers are produced in different inflorescences, but do not open at exactly the same time, so that cross-pollination predominates. Nevertheless, maize is self-fertile and produces seed if it is artificially self-pollinated. It is therefore possible to produce so-called *inbred lines*, corresponding genetically to pure lines, though they lack vigour and yield poorly. In its simplest form, seed of an F1 hybrid cultivar is produced by crossing two of these inbred lines. The hybrid plants are vigorous, high-yielding and completely uniform but, if the grain harvested from these hybrid plants is sown as seed, uniformity is lost and the plants of the next generation show the various characters of the original inbred lines in different combinations. Some of the vigour, too, is lost in subsequent generations. Farmers who grow F1 hybrid cultivars should never sow seed harvested from their grain crops, but obtain hybrid seed each year from specialist seed growers.

A plant breeder may produce a cultivar especially suited to the conditions of one particular region. The rainfall, temperature, day-length, soil type, diseases, pests, etc. which are prevalent in the region are known, and in these conditions the cultivar gives high yields. For such narrowly defined growing conditions, uniformity within the cultivar is important, so that each individual plant will contribute equally to the harvest. Some regions, however, are not uniform; they vary from place to place because of altitude, aspect or soil, or they vary from one year to another, notably in rainfall. In such a region, a uniform cultivar may fail completely in the places or in the years to which it is not adapted. A cultivar which is a mixture, however, is adaptable; it may not yield very well in good conditions, but in adverse situations or seasons some of the plants will grow well enough to produce a yield that is welcome in the circumstances.

So, although uniformity is important for maximum yields in reliable arable districts, there is room in a country's repertoire of cultivars for some that are not uniform. To meet this need *synthetic cultivars* (sometimes called *composites*) have been developed and have been particularly successful in maize, perhaps less so in cotton. In principle, a synthetic cultivar is produced by mixing together seeds of a number (say, 10 to 15) of pure or inbred lines. The original mixing is done by the breeder, and the mixed stock is then multiplied over a few generations under control. During this multiplication the plants inter-cross to an extent depending on the breeding system, and the distinctions between the constituent lines break down. At the end of the controlled multiplication programme, the seed is sold to farmers who then grow it for several generations. A synthetic cultivar yields less than a F1 hybrid cultivar, but more than unimproved native varieties. It is much less troublesome to produce than a hybrid cultivar, and the farmer does not have to purchase new seed every year.

From the foregoing it is clear that there is no simple comprehensive definition of a cultivar. The international code of nomenclature of cultivated plants defines a cultivar as

> an assemblage of cultivated plants which is clearly distinguished by any characters and which, when reproduced retains its distinguishing characters.

This puts the emphasis on reproducibility. Breeding systems vary, and in natural conditions the distinguishing characters between cultivars may be lost. The task of the seed producer is to control or modify the breeding system so that they are indeed retained.

Deterioration of seed stocks

Assuming that a stock of seed when it leaves the plant breeding station is of high quality in every aspect, deterioration may occur in the subsequent multiplication.

Loss of quality can happen in one season. Poor germination and vigour, disease, contamination with weeds, and admixture with other kinds of crop seeds can follow catastrophically from bad weather at harvest time, poor husbandry or an accident in the seed store. Quality, however, does not always deteriorate suddenly; it can worsen gradually over a number of generations. This can be minimized by good husbandry, by roguing, by the many precautions taken for this purpose, and by thorough cleaning of seed after harvest. The ordinary farmer takes no special precautions; he does not possess the equipment necessary for the proper cleaning of harvested seed, and quality can deteriorate rapidly when he saves his own.

The incidence of seed-borne diseases in a stock can increase each generation, particularly in climatic conditions that favour disease. Under a controlled system this can be checked by multiplying only in regions known to yield healthy seed, by eliminating infected seed crops, and by treating the harvested seed. Seed treatment is a possibility for the ordinary farmer, but it is not always practised.

A cultivar may appear to lose its resistance to a particular disease. This, however, is not due to any deterioration in the cultivar, but to a genetic change in the pathogen. For example, black rust is caused by a species of fungus, but the species includes many races, corresponding to plant cultivars and differing in their ability to parasitize wheat plants. A new cultivar of wheat may be resistant to the existing races, but after a few years a new race of the fungus may develop which is capable of attacking the new cultivar. In consequence, the cultivar appears to lose its powers of resistance, though in fact it has not changed.

Deterioration can occur in the genetic constitution of the cultivar, due mainly to outcrossing. The risk of this happening depends on the extent of cross-pollination in the breeding system of the cultivar. It is rare in wheat and barley which are almost entirely self-pollinated, occurs to some extent in rye and sorghum, and is common in the cross-pollinated grasses. Multiplication systems and isolation requirements for different crop species are planned according to the risks involved.

In pure lines, the effect of cross-pollination is not limited to the introduction of foreign genes. The seed produced will give rise to a plant with hybrid vigour, which will in its turn produce more seeds than its pure-bred neighbours. The proportion of rogues in the stock is thereby multiplied.

More insidious is the danger of genetic change within a cross-pollinated cultivar, even with no out-pollination. A cultivar of this type is a non-uniform population of heterozygous plants susceptible to selection. If it includes plants of different height, for example, and the tallest plants are removed over several generations, the average height decreases; similarly, if all the short plants are removed, the average height is increased. This phenomenon is known as *genetic shift* and can occur when a cultivar is multiplied out of its original environment.

This was well shown in an international experiment in which seed of Grimm, a late prostrate cultivar of lucerne bred in Canada, was multiplied over three generations in Cyprus and the seed harvested there sown for trial in England. Within two generations the stock had changed to an earlier, more erect, higher yielding type of lucerne. Genetic shift does not necessarily occur in every case, but the possibility has been established. To prevent it happening, basic seed of a cross-pollinated cultivar is always produced in the environment to which it is adapted, and the number of subsequent generations of multiplication are strictly limited.

To take a more specific example, in snap beans *(Phaseolus vulgaris)* a rogue occurs with a flat fibrous pod which is unsuitable for canning. The rogue is also early-flowering and, because the pod is more mature and fibrous at harvest, it is easier to thresh than normal pods. The result is that the proportion of seeds of this rogue tends to increase each generation.

Genetic changes may also be due to irregularities in the number of chromosomes. Meiosis does not always work perfectly and occasionally gametes are produced with the wrong number, or with broken pieces, of chromosomes. There is also the possibility of a mutation in one of the genes.

Production of seed for the farmer

When a farmer buys seed, he sows it in the year of purchase and then, for a number of years, sows seed which he has saved himself from his previous harvest. The number of generations over which he can do this depends partly on the purity and genetic stability of the seed he buys, and partly on his own success in preventing deterioration. He does not necessarily buy all his seed requirements for one year and then no seed for several years in succession, but may buy part of his requirements every year.

The seed which he buys may be produced under a controlled system, or without any control. Historically, free marketing of seed came first and controlled systems, voluntary and official, came later, primarily to give purchasers an assurance of cultivar purity. The two systems are not mutually exclusive, and both can exist together in the same country.

In an uncontrolled regime, seed is not always harvested from a crop grown for that purpose. The decision to harvest for seed may be influenced by the weather or by the market price, and the whole crop might not be used in this way. The produce of a cereal crop might be used partly for grain and partly for seed; part of a grass crop might be cut for forage or grazed, and the remainder left to set seed. Herbage seed may be harvested from natural grassland or from a sward that had been originally sown with such seed. This practice gave rise by a process of selection to local cultivars, which were not produced by plant breeders, but were evolved naturally and were typical of their region of origin. Without control, there is no check on the authenticity of the seed sown or the purity of the seed harvested, and no official inspection of the growing crop.

In some countries the cultivar control system that has been set up is voluntary. Control is exercised by an organization which may be an association of plant breeders, seedsmen and farmers, or may be a government agency. The main purpose of this organization is to certify the cultivar purity of seed lots, and this is indicated by labelling and sealing the bags of seed. Participation in the scheme is voluntary; plant breeders, seedsmen and seed-producing farmers are free to choose whether or not their seed is to be multiplied under the scheme. Farmers who do not produce seed are able to choose between buying seed certified by the organization, or uncontrolled seed of doubtful authenticity which is freely marketed at a lower price.

In other countries, control is compulsory and seed may not be sold unless it has been certified by a government agency. Nevertheless, farmers are free to save and sow their own seed—the control applies only to marketing.

A system of partial control may be followed where there is no certification scheme. The breeder produces a large quantity of seed each year and distributes it to a limited number of experienced growers with a wide geographical spread. These growers then sell the next generation of seed to neighbouring farmers, but are not subject to crop inspection or any other form of control. Alternatively, the breeder may multiply his stock with the aid of contract growers and then market the produce through the best available channels. These systems can be effective for stable cultivars such as pure lines, but are not suitable for cultivars which are unstable or require isolation.

Controlled multiplication

The system followed in this book is the one agreed by the national certification authorities which participate in the OECD seed schemes. Some national schemes differ in detail, but not fundamentally, the differences being mainly in nomenclature; descriptions such as foundation, registered, elite, and approved stock are used, but not always with the same meaning. The OECD scheme seeks to avoid confusion by using terms that are precisely defined.

The principles of the OECD scheme can be summarized as follows:

1. Only cultivars officially recognized as distinct and of acceptable value are admitted.
2. All seed is directly related through one or more generations to Basic Seed.
3. All stages of multiplication are subject to pre-control, post-control and field inspection (see chapter 12).
4. Close co-operation is maintained betv een breeder and certification authority.

Basic seed

The key to the system is *basic seed*. This is seed produced by the breeder of the cultivar and is intended for the production of certified seed. It may not have been actually produced by the breeder, but it has certainly been produced under his responsibility and under the control of the certification authority, and is therefore acceptable as genuine. The quantity of basic seed of a cultivar produced in a year may be anything from a few hundred kilograms of herbage seed to at least as many tons of cereal or pulse seed. As a precaution against loss of the crop, it should not all be produced on one farm, and anything up to half of it should be stored as a reserve in case of catastrophic crop losses on a national scale. Apart from this, all basic seed should be used for the production of certified seed in order to ensure the best possible utilization of valuable genetic material.

Basic seed is derived from the breeder's *parental material.* This is the small population of plants which the breeder himself grows from year to year to maintain the cultivar. Pollination is strictly controlled and the plants may be enclosed under glass or polythene. To ensure that no genetic change occurs, there is positive selection of plants showing the characteristics of the cultivar; the material is examined meticulously and any doubtful plants removed. From this parent material a few hundred seeds or plants are taken each year for multiplication under strict observation. This is usually referred to as *breeder's seed* and is multiplied over several generations to produce basic seed, the penultimate generation being known as *pre-basic seed.* Alternatively, a quantity of the original seed may be kept in conditioned storage and a few seeds taken out each year to serve as breeder's seed.

Certified seed

Basic seed is sold to seedsmen, co-operatives or individual growers, and so passes out of the control of the breeder. Further multiplication is under a certification scheme and the seed produced is designated *certified seed.* Certified seed is intended either for the production of another generation of certified seed or for sale to farmers to sow for their own use. The seed produced directly from basic seed is designated Certified Seed, 1st Generation, and later generations of certified seed are numbered in sequence. A certification authority may place a limit on the number of generations that can be produced, thus ensuring an injection of basic seed into the scheme every year.

The system can be set out diagrammatically as follows:

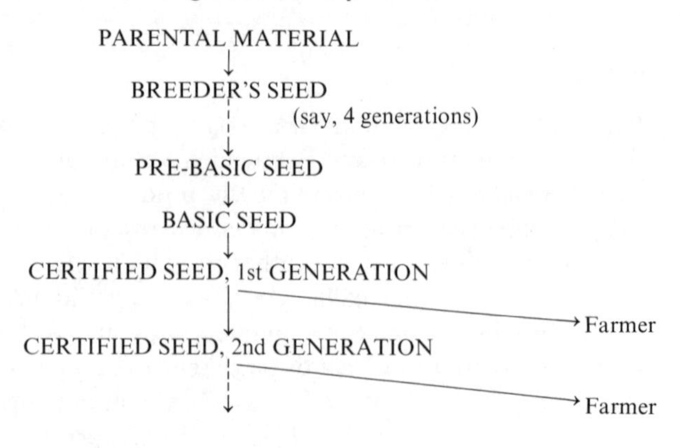

In applying this system to F1 hybrid cultivars, the key is that certified seed is sold to farmers for the production of grain, and basic seed is sown for the production of certified seed. The inbred lines are maintained by their breeders.

In the case of a single cross hybrid cultivar, basic seed is the seed of two inbred lines thus:

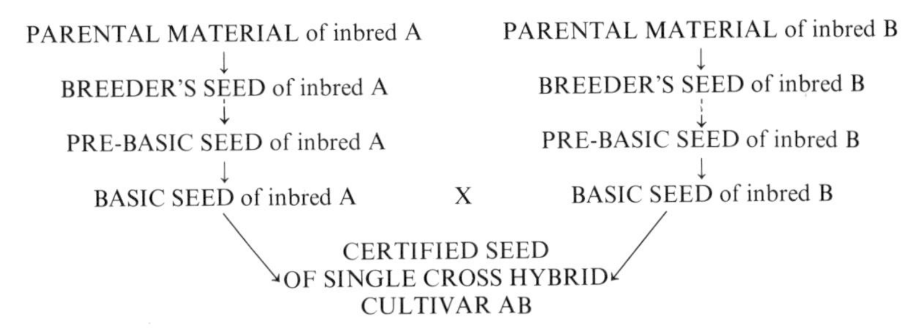

For the production of a double cross hybrid, four inbred lines are involved and basic seed is seed of two single cross hybrids, thus:

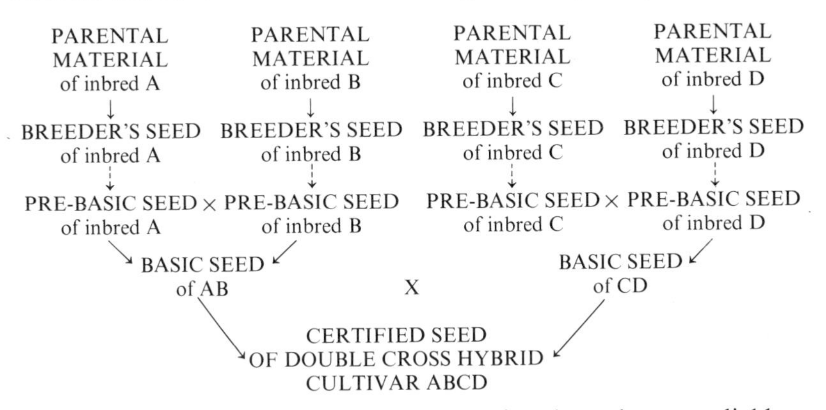

The success of hybrid maize seed production depends on a reliable supply of seed of many inbred lines. These inbred lines, sometimes described as *foundation* or *elite seed*, are maintained by official and commercial plant breeding stations, by co-operatives or by individuals. The marketing of inbred seed may be direct to growers or through a co-operative association. The production of hybrid seed may therefore involve a number of organizations, e.g. for a double cross, the inbred lines may be supplied by four different companies, the single cross hybrids by two more, and the final double cross by a seventh.

Planning

It takes at least four years to multiply seed from the breeder's parental material to the stage of certification. The actual time depends upon the quantity of breeder's seed, the amount of certified seed required, and the multiplication factor. The requirement of farmers for certified seed has therefore to be anticipated, and a multiplication programme planned several years in advance. Seed cannot be manufactured overnight to satisfy an unexpected market demand.

Requirements depend on the frequency of renewal of seed on the farms; this is influenced by the genetic stability of the cultivar, the skill of the farmer, and the nature of the end product. The production of certified seed is expensive in terms of money, manpower and resources, and should be limited to what is necessary. Farmers who buy it should be encouraged to use it efficiently by protecting the seed they save on their own farms against contamination, disease and loss of viability. They are advised to buy seed of an F1 hybrid cultivar every year, but seed of a synthetic cultivar needs to be renewed only after three to five years, and seed of a pure line is frequently maintained on a farm for six years or more. The certified seed requirement in any one year may therefore be no more than enough to sow about 20 percent of the total area.

In the case of cotton, it is particularly important to ensure uniformity in the lint characters of the ginned fibre. This necessitates a high standard of cultivar purity in all crops, and a farmer is required to sow fresh seed every year and send all the cotton picked from his field to the ginnery. In the following year, seed is issued to him free of charge, and the cost deducted from the price of picked cotton delivered to the ginnery. Nevertheless, there is always the possibility that some cotton may be retained for village weaving and contaminate the following year's crop.

Seed multiplication factor

The seed multiplication factor is the number of kilograms of good seed harvested from each kilogram sown.

Obviously it must differ from one crop species to another, but even within a species there is a wide range of variation between regions and between individual growers. In general it depends on the fertility of the soil, the standard of husbandry, the weather, and other agronomic factors. More particularly, it is influenced by the skill of individual farmers in sowing the minimum amount of seed and harvesting as much as possible in good condition. In maize, more seed is obtained from a double cross than from a single cross, because the female parent is itself a hybrid and therefore more productive. The following figures indicate the range of factors that are achieved in practice:

	Seed Multiplication Factor
Wheat	16 to 23
Barley	13 to 20
Rice	20 to 30
Maize	45 to 250
Groundnut	10 to 33
Soya	15 to 40
Cotton	8 to 40
Lucerne	40 to 800
Grasses	20 to 100

FURTHER READING

Department of Agriculture (1961), *Yearbook of Agriculture*, Washington.

11

TESTING FOR CULTIVAR AUTHENTICITY AND PURITY

THE PRODUCTION OF SEED FOR SALE TO THE ORDINARY FARMER INVOLVES taking a small quantity of seed from the breeder and multiplying it over several generations. Breeder's seed is true to cultivar and 100 percent pure, but the stock may deteriorate during the multiplication process. Deterioration may be due to

genetic factors, such as cross-pollination
contamination with seeds of other cultivars, as may occur in the processing plant
substitution of another cultivar due to wrong labelling

Measures designed to prevent such deterioration are taken during growing, harvesting, drying and processing, and successful seed production depends on strict observance of these precautions. However, accidents can happen and a cultivar check is carried out on every crop or seed lot. The purpose of this check is to determine

whether or not the stock is of the cultivar it is believed to be, and
if it is on the whole true to cultivar, the number and (if possible) the identity of any plants of other cultivars that may be present, i.e. the cultivar purity.

This requires the examination of a sample either of plants or of seeds, or of both. As explained in chapter 10, the cultivar characteristics which are of interest to the farmer or consumer are often not apparent in these examinations, and cultivars have to be identified by associated marker characters. When a cultivar is accepted into a certification scheme, a detailed description is prepared of the plants and seeds, and this description emphasizes the characters by which the cultivar can be distinguished from others; in an examination of plants or seeds, these characters in particular are checked. It is rare for a cultivar to be identified by one character; usually it is identified by a particular combination of characters.

As a general rule, a cultivar cannot be identified by examination of the seed; the seed either carries no diagnostic features or not enough for complete identification. In practice, the examination may take the form of

field inspection of a crop from which seed is to be harvested,
examination of plants in a control plot sown with a sample of the seed being multiplied, or,
a laboratory test of seeds or seedlings.

It is convenient to classify diagnostic characters according to the kind of examination in which they can be seen.

These examinations are not designed to answer the question—to which cultivar do these plants or seeds belong? That is a wide question which cannot be answered by a routine test, because it involves observation of *all* the characters used in the species for distinguishing cultivars and the availability of many cultivars (ideally all of them!) for comparison. The question to be answered in routine tests is more limited—do these plants or seeds on the whole belong to cultivar X, and if so, what is the percentage purity? This requires observation only of the characters used to distinguish cultivar X and, for comparison, authentic seed of cultivar X only.

An examination may be carried out by the breeder, merchant or grower responsible for multiplying a seed stock, or officially by a certification authority to establish that the material complies with the prescribed standards.

Field inspection

A field inspection is usually carried out after inflorescences have developed in the crop, at which time cultivar differences are most obvious. Characters which are apparent for a short time and might disappear or change before inspection are not suitable, e.g. angle of growth or colour of anthers. As for roguing, characters which can be used must be visible to an inspector, without using a lens, while walking slowly through the crop. Some examples are:

Wheat—length and density of ear, and presence or absence of awns (figure 11.1)
Barley—number of rows in the ear, ear density, length of awns, and pigment on the auricles of the flag leaf
Oats—type of panicle

Figure 11.1 Ears of two wheat cultivars showing differences in the length and number of awns (Department of Agriculture and Fisheries for Scotland).

Maize—colours of silks and husks, number of nodes
Rice—presence or absence of awns
Peas—flower colour
Soya—colour of hairs on stems and of flowers

Height and maturity are characters that can be used to some extent in all species. A tall impurity can be seen in a short cultivar, but a short plant would not be noticed in a tall crop; similarly, an early-flowering plant can be seen in a late-flowering crop, but not vice versa.

A routine field inspection is not a reliable method of determining cultivar authenticity because in the conditions of the examination it is impossible to differentiate between closely related cultivars and no control is available for comparison. Assuming, however, that the crop is of the correct cultivar, the inspection does provide an estimate of the incidence of obvious impurities and, if these are excessive, the crop is not worth harvesting for seed. It is not practicable to examine the whole crop in detail, and the inspection procedure is in three parts:

confirmation of cultivar
a general view of the crop
detailed examination of a random sample of plants

To confirm as far as possible that the crop is of the correct cultivar, the inspector makes a detailed examination, using a lens if necessary, of about 100 plants taken at random over the field.

To obtain a general view, the inspector walks round the crop, noting the neighbouring crops and how far they are separated, possible sources of foreign pollen and the degree of isolation, gateways and corners which are particularly liable to weed growth and other contamination. This is followed by a walk across the field, following a zig-zag path, but in such a way that every row of the crop is crossed. During this walk, areas of poor growth, excessive weed contamination and lodging are noted; obvious off-types, rogues and noxious weeds are identified. If the crop is below shoulder height, it is helpful to bend down from time to time and observe the crop at ear-top level; from this angle, tall impurities stand out clearly, particularly against the background of the sky.

This general view of the crop is interrupted by the detailed examination of small areas, or quadrats, taken at random over the field. Within each quadrat a meticulous search is made for off-types and rogues, a lens being used to confirm doubtful cases, and their number is counted. If the habit of growth permits, the number of plants or inflorescences in a metre length of row within the quadrat is counted, and the distance between the rows is measured. The number of plants or inflorescences within the quadrat is estimated thereby, and on this basis the percentage of impurities can be calculated. At least ten quadrats are examined and (if possible) each should include about 10 000 plants or inflorescences; this gives an accuracy of 0.01 percent of cultivar purity, assuming that all impurities are recognizable. However, in widely spaced crops such as maize and sorghum, the number of plants examined in each quadrat may be as low as 100. In herbage crops no attempt is made to estimate the population; each quadrat is of a standard size, say 20 square metres, and the number of impurities is expressed as the number per quadrat.

Impurities such as other species and noxious weeds can be counted at the same time as off-types and rogues, and their incidence calculated. However, the acceptable level of these contaminants is lower, and it is necessary to take into account observations made outside the normal quadrats.

Control plots

When a sample of the seed that is to be multiplied is sown in a field plot, the plants are available for examination throughout the growing season, in contrast to the brief examination at one stage of development that is possible in a field inspection. Furthermore, the work is organized so that time and facilities are available for examining minute structures and measuring variable characters. Cultivar authenticity can be assessed by comparison with a neighbouring plot sown with seed known to be genuine. As an insurance against failure, the seed sample should be sown in at least two widely separated replicate plots.

The seed is sown in conditions conducive to good growth and the expression of cultivar characteristics. This requires a fertile soil, control of pests and diseases, in a climate and season to which the cultivar is adapted. In the case of biennial and perennial species, the plot may have to be continued into a second growing season. If it is required to expedite flowering or to make an examination out of season, it is possible to grow the plants in a glasshouse or growth cabinet, but this is generally too costly for routine examinations, and only a small number of plants can be grown.

The space available to a plant can influence its habit of growth and, in consequence, might modify the expression of cultivar differences. For example, in the crowded conditions of a farm crop, a wheat plant usually produces up to three stems but, if given sufficient space, it will grow into a spreading plant with numerous stems and irregular ripening. In plot tests, therefore, spacing is important, and the general rule is that each plant is given slightly more space than is normal in a field crop, so that the plants can be separated from each other and counted accurately. The plants are grown in rows, close together for small-grained cereals, but more widely spaced for maize and root crops. In some species, however, it may be necessary to give each plant considerably more space than in a field crop, in order to permit the full expression of cultivar differences. In such cases, the plants are grown, not in rows, but as spaced plants. Typical numbers of plants per metre length of row are:

small-grained cereals	60
soya	30
Vicia faba	10

A standard of cultivar purity of 99.9 percent means that one impurity is permitted in 1000 plants. To obtain a reliable estimate of purity at this level, about 10 000 plants should be grown. In the case of small-grained cereals, this requires a row length of about 170 metres, which can be accommodated in a total plot area of 40 square metres. For soya, *Vicia faba* and other pulse crops, the plots should be even larger. It has to be recognized, however, that land, staff and other requirements may not be available for plots of this size and, in practice, smaller plots are often used; for root crops, the number of plants may be only 400, or for maize 100.

Grasses are grown as spaced plants or in rows, depending on what is required. As a rule, grass cultivars are distinguished, not by specific marker characters, but by growth characters such as habit, date of ear emergence and length of stem. The trueness to cultivar of a sample can be judged only if each plant is given sufficient space for these characters to develop. This requires a spacing of at least 600 mm, and usually it is not practicable to grow more than 100 plants. However, the most consistent difference between cultivars lies in the date of ear emergence and, if only a rough indication of trueness to cultivar is required, this can be estimated in a thinly sown row of about 15 metres.

In the row method, the seed is sown thinly along the row, and surplus plants removed later if necessary. For spaced plants, single seeds are planted in boxes or small pots, and the plants transplanted to the field plot when they have grown sufficiently.

During the growing season, the plot is examined at frequent intervals, even daily at critical periods, always comparing it with the control plot sown with authentic seed. Plants identified as belonging to the wrong cultivar are marked. At the stage when characters not evident to the unaided eyes are expected to develop, a number of plants is uprooted for more intensive scrutiny. This may be with a hand lens on the spot, or by a binocular microscope in a field laboratory. These observations continue right up to the stage of harvest or later. In the case of root crops, the roots are lifted, cleaned and laid out in lines to display their size, shape and colour.

Characters which can be observed in all species include habit of growth, angle at which leaves are held, and date of ear emergence or flowering. Some examples of characters which can be observed in particular species when grown in rows, follow.

Wheat—Glume colour. Thickness of stem wall (figure 11.2). Hairs on rachis. Lemma shape.
Oats—Hairs on leaf and node. Hairs at base of grain (figure 11.3).
Maize—Colour of seedling leaf sheath. Hairs on leaf sheath below ear.
Barley—Length of first rachis internode. Anthocyanin on lemma. Hairs on rachilla (figure 11.4). Leaf sheath hairs. Shape of collar at base of ear.
Rice—Colour of stigma. Length of panicle.

When a character has a numerical value, e.g. length of stem, angle of leaf, or date of flowering, the mean and standard deviation are calculated and compared with corresponding values from the control plot.

Laboratory tests

Cultivar differences which are expressed in the seeds or seedlings can be observed in laboratory tests. An authentic sample should always be available for comparison. The sample usually contains 400–1000 seeds, giving an accuracy of 0.1 percent. The extent to which laboratory tests are used varies in different countries, but it is claimed

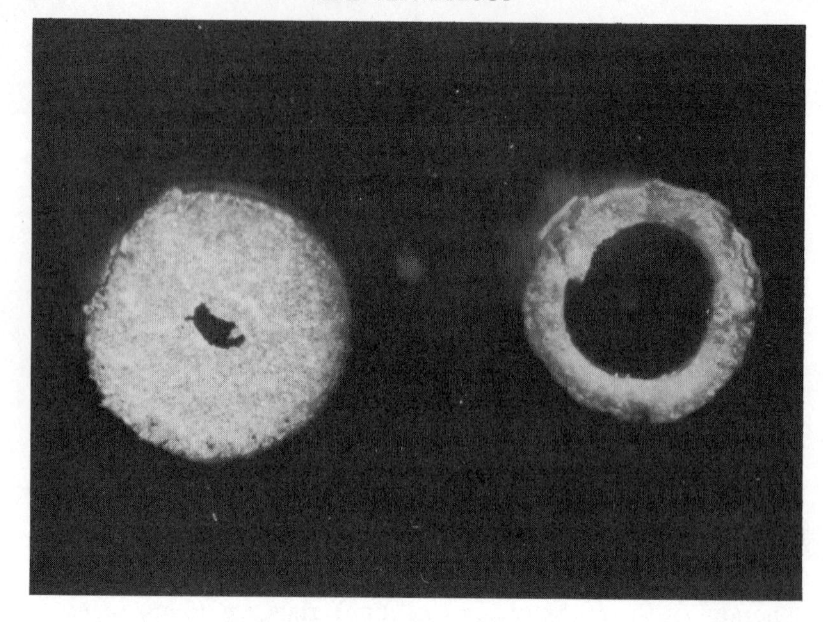

Figure 11.2 Cross-sections of the stems of two wheat cultivars showing difference in the thickness of the wall (Department of Agriculture and Fisheries for Scotland).

Figure 11.3 Grains of two oat cultivars showing difference in length of basal hairs (Department of Agriculture and Fisheries for Scotland).

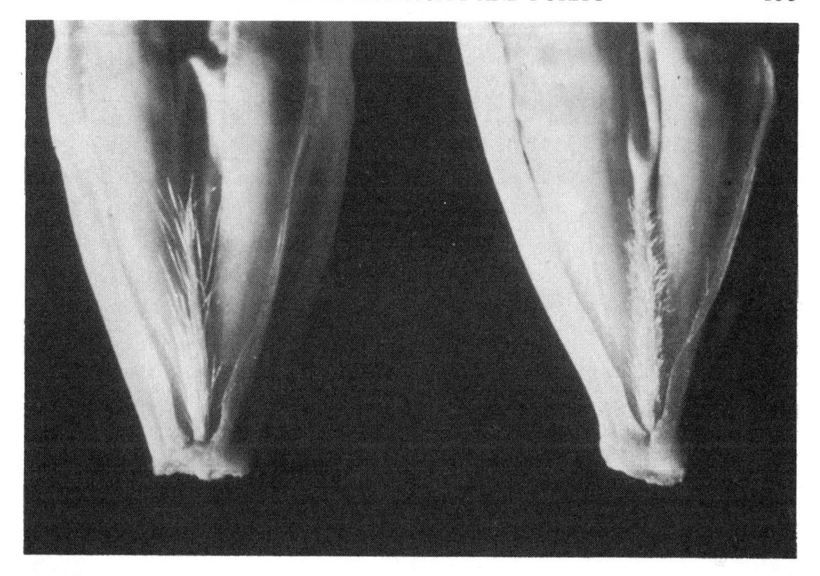

Figure 11.4 Grains of two barley cultivars showing differences in length and density of hairs on the rachilla (Department of Agriculture and Fisheries for Scotland).

that all the wheat cultivars grown in New Zealand can be identified in this way.

Examination of dry seeds

Differences between cultivars may be in structure, size, or colour of their seeds. When the "seed" is a true seed in the botanical sense, there is little scope for variation; differences, if any, are usually limited to colour and size. There are however, slight differences in seed shape in lucerne, while pea cultivars differ in hilum shape and in the wrinkling of the testa.

In many crop species the true seed is enclosed within a fruit, which may in turn be surrounded by other organs and appendages (see page 31), and these provide much more scope for slight but significant differences between cultivars. These differences have been most fully exploited in cereals, and some examples follow of characters that are commonly used in cultivar identification:

Wheat—Colour, shape and size of caryopsis. Frequency and length of hairs at apex of caryopsis.
Barley—Length of hairs on the rachilla (figure 11.4). Presence or absence of teeth on the nerves of the lemma (figure 11.5). Colour of aleurone layer.
Oats—Colour of lemma. Presence or absence of hairs on the rachilla.
Rice—Colour of lemma.
Maize—Colour, size and indentation of caryopsis.

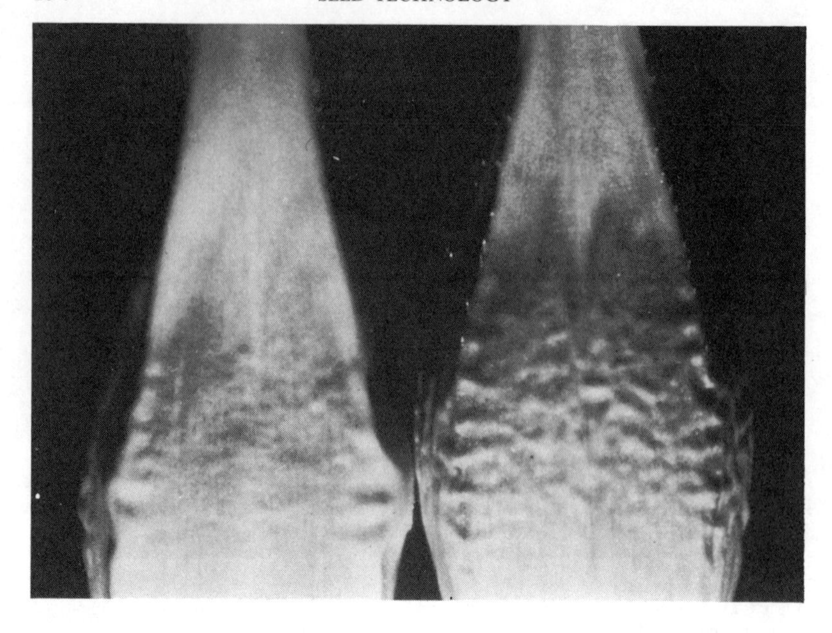

Figure 11.5 Grains of two barley cultivars showing presence (right) and absence (left) of spines on the lateral nerves of the lemma (Department of Agriculture and Fisheries, Scotland).

Seed characters may be lost during processing, e.g. the teeth on the barley lemma and the hairs at the base of the oat lemma may be rubbed off. These characters can, however, be observed in inflorescences harvested in a plot examination.

Chemical tests

It is known that slight chemical differences exist between seeds of different cultivars, and techniques which can demonstrate the differences have been developed. The differences are not so much in the presence or absence of particular substances as in the proportions of the various substances present. This variable pattern can be most effectively demonstrated by chromatography; so far it has not been used to any extent, but there are possibilitites for the future in confirming or otherwise the identity of the cultivar rather than in estimating cultivar purity. In Australia, 13 patterns have been found in barley enabling most (but not all) of the cultivars grown there to be distinguished. In England out of 29 wheat cultivars tested, all but three showed distinguishable patterns.

The only chemical test that is commonly used today is the relatively simple phenol test for distinguishing wheat cultivars. Seeds placed on a blotter soaked with 1 percent phenol turn brown, ranging from a light to a dark shade. A cultivar cannot be identified precisely, but can be allotted to a group according to the shade of colour. Some cultivars show various shades in different seeds, but the proportion of each is typical of the cultivar or its group.

Examination of seedlings

Cultivar differences in seedlings can be observed by germinating seeds in the laboratory and allowing the seedlings to grow for no longer than is required for a germination test.

The most useful character is colour. In cereals the coleoptile may be red or colourless, and in soya the hypocotyl may be purple or green. In beet the hypocotyl may be dark green or yellow-green or, if grown in the dark, may be seen to develop a pink colour. Turnip seedlings, if grown in the dark, have yellow or white cotyledons corresponding to the colour of the root in the adult plant. Colour differences which are not normally visible may appear under ultra-violet light, e.g. in the roots of beet seedlings:

Sugar beet—blue-white fluorescence
Sugar/fodder beet—blue-green fluorescence
Fodder beet—green fluorescence

Pubescence on the seedlings may provide identification clues, as in oats and soya. In soya the hairs differ in sharpness and in the angle at which they lie.

Barley cultivars differ strikingly in their reaction to the insecticide DDT. A spray of the insecticide is lethal to the seedlings of some cultivars, but not of others and this can be used as a distinguishing test.

In many crop species, plant breeders have produced new cultivars which are resistant to particular plant diseases. Such cultivars, however, are not completely resistant to the disease, but only to certain strains of the pathogen. It may, therefore, be possible to distinguish between cultivars by inoculating seedlings with strains of the pathogen to which they are known to be susceptible or immune. Such tests, indeed, are the only known means of distinguishing between certain cultivars of lettuce. In this case the pathogen is the fungus *Bremia lactucae* and the tests are possible only if a plant pathology laboratory maintains pure cultures of the various strains.

Number of chromosomes

Within a species, as a general rule, all plants have the same number of chromosomes. For the species there is a basic set, and in each cell there are two sets. In some species, however, cultivars may have different chromosome numbers, made up of more than two sets. This increase may have happened by chance, or may have been deliberately induced by a plant breeder. In perennial ryegrass and red clover, the basic set consists of 7 chromosomes. Most cultivars have two sets, making 14 (called diploids), but some have four sets, making 28 (called tetraploids). In beet the basic number is 9 but, in addition to diploids and tetraploids, there are cultivars with three sets, and these are called triploids.

When a seed germinates, the cells in the root tip of the seedling divide very rapidly for a time; at this stage, using appropriate techniques, it is possible to see the chromosomes with a powerful microscope and to count them. The number determined in this way gives a clue to the cultivar, and in some seed-testing stations this test is performed as a routine.

FURTHER READING

Carson, G. P. and Horne, F. R. (1962), "The identification of barley varieties" in *Barley and Malt*, edited by A. M. Cook, Academic Press, N.Y.

ISTA (1964), "Varietal purity examination," *Proceedings of the International Seed Testing Association*, Vol. 29, No. 4.

Ulvinen, O. and others, "Testing for genuineness of cultivar," Part of *Nordic Handbook on Seed Testing*. English version issued by ISTA.

ISTA (1975). "Report of the Variety Committee 1971–74," *Seed Science and Technology*, Vol. 3, No. 1.

12

CERTIFICATION

IT WAS EXPLAINED IN CHAPTER ONE THAT SEED QUALITY IS MADE UP OF ten components and that all of them can be assessed by a laboratory test except one, and that one is cultivar purity. It is a relatively simple matter to ensure that the seed sold to a farmer is of high quality in respect of the attributes that can be measured in a laboratory. When a seed lot is packaged and the bags sealed and labelled, a sample is drawn for test, then, provided the lot is not stored overlong or in poor conditions, the farmer can be assured that the quality of the seed he receives is up to the standard shown in the test.

A high standard of cultivar purity, however, can be assured only by controlling the production of seed. Absolute control over all the operations involved is not possible, but control is exercised as far as is practicable in two ways:

1. By ensuring that seed is multiplied and processed in such a way as to minimize the risks of mechanical and genetic contamination.
2. By setting standards and checking each seed stock against these standards at a stage in which cultivars can be distinguished. This cannot be done by examination of the seeds, but by examining the plants from which the seed is to be harvested.

To exercise this control, a certification authority is established. This authority may be a government department or agency, or it may be a voluntary association of breeders, seed merchants and growers. The authority prescribes the conditions under which seed is to be produced and the standards which must be attained. A seed lot which is produced under these conditions and which attains these standards is certified as such by the authority, and this is indicated by labels and seals attached to the bags delivered to farmers. Although certification is primarily for the purpose of assuring cultivar authenticity and purity, it is convenient to combine this with an assurance of high quality in such attributes as analytical purity, germination capacity and health, which can be assessed in a laboratory.

A certification authority cannot certify cultivar purity unless its officers are familiar with the cultivar, and the process of certification is possible only if the cultivar is distinct, uniform and stable up to the number of generations to which it is to be certified. Furthermore, certification involves the expenditure of much effort and money, and this can be justified only if the cultivar is of sufficient value to the farming industry.

The certification of a seed lot involves a complex pattern of procedures. The technical operations require special skills and experience and have to be completed in accordance with a predetermined seasonal timetable. Linked to the technical operations is an enormous amount of office work, mainly checking applications and issuing labels, which requires meticulous attention to details. This is not the place for a detailed account of all that has to be done, but what follows is a general guide to the procedures, indicating the principles involved. The actual details vary from one country to another, and according to the crop species. The account falls under three heads:

The conditions that must be met in producing the seed lot to be certified.
The standards to be attained in the crop and in the seed that is ultimately certified.
The procedures to be followed to ensure as far as possible that these conditions and standards are satisfied.

Conditions

Multiplication system

A controlled multiplication system is prescribed under which all certified seed is related directly, through one or more generations, to authentic basic seed of the cultivar. Basic seed is produced by, or under the responsibility of, the breeder and each lot is approved, sealed and labelled by the certification authority. The main factors determining the standard of certified seed are the cultivar purity of basic seed and the number of generations from it. Certification schemes may limit the number of generations, allowing more in self-pollinated than in cross-pollinated species.

There are good biological reasons for restriction of generations in the case of cross-pollinated species. Accidental out-pollination may occur, cultivars are liable to genetic shift, and the consequences may not be evident in a field inspection or in the time and space available for post-control plots. The actual number of generations permissible depends on the breeding system and should be based on local experience.

By contrast, self-pollinating species such as barley and wheat are stable, and cultivars can remain genetically unchanged over an unlimited number of generations. Deterioration, when it occurs, is due to disease or, more frequently, to mechanical contamination with seeds of other cultivars and species due to human carelessness or incompetence. Experience has shown that on the average this happens in less than five generations, and its occurrence is readily apparent in field inspections and post-control plots. The argument for restriction of generations in crop species of this kind is that specialists with the necessary skill and experience are limited, and their efforts should be concentrated on pre-basic and basic seed. If basic seed is of high purity, serious deterioration, though possible, is improbable within two or three generations, and inspections of crops can therefore be less thorough or done by people with less experience.

Eligibility of cultivars

The certification authority publishes a list of the cultivars, the seed of which it is prepared to certify. To be listed, a cultivar must satisfy the authority as to its distinctness, uniformity, stability and agricultural value. For each listed cultivar, a maintainer is nominated who is responsible for the production of basic seed. Normally the maintainer is the original breeder of the cultivar or his agent. A condition of acceptance is that all basic seed be made available for the production of certified seed. At the time of acceptance, a large sample of seed is put into long-term storage by the certification authority, and this is used as a check against seed of the cultivar produced by the breeder in later years.

The seed to be sown

For growers other than the breeder, this must be either basic seed or certified seed of a generation acceptable for further multiplication.

The ground on which the seed crop is to be grown

To prevent contamination by seeds left in the ground from previous crops, there are rules (where practicable) as to how many years must have elapsed since a crop of the same species or of a similar type was grown on the same land. Seed of the same cultivar, however, may be produced in successive years provided the previous crop was of certified standard. If the crop is liable to be cross-pollinated, there are rules about minimum isolation distances.

The seed crop

The crop must show vigorous healthy growth, displaying the characteristic features of the cultivar, and must not be excessively weedy. If it is of a species from which more than one seed harvest may be taken, e.g. perennial forage crops, the number of harvests may be limited.

Adequacy of harvesting and processing arrangements

The certification authority has to be satisfied that adequate provision is made for harvesting, drying, cleaning and storing the seed in order to ensure that these operations are performed effectively. This involves not only the type of machinery and equipment used, but also its capacity and its manning.

Maintenance of identity

Throughout the whole series of operations, the stock must be kept separate from all other stocks and must be identifiable. This applies to the seed that is sown, to the crop in the field, and to the harvested seed on the farm, in transit, in the processing plant, and in store.

Standards

Except in the case of F1 hybrid cultivars, a farmer expects, after an initial purchase, to save his own seed for a number of years, and the certification standards are high enough to allow for some of the deterioration that inevitably occurs during this period. These standards refer to the growing crop and to the seed lot to be certified.

The growing crop

The most important crop standard concerns cultivar purity, indeed the whole certification system is based on the need for an examination of growing plants. The determination of cultivar purity requires an estimate of the number of rogues or off-types present in the crop. A *rogue* is a plant which obviously belongs to another cultivar, while an *off-type* is less distinctive and may be a genetic variant. In the case of cereal and pulse crops in which the number of plants or ears can be counted, the standard is expressed as the maximum permissible percentage of plants or ears of rogues and off-types. In forage crops, in which the plant population is denser and individual plants cannot be counted, the standard is expressed as the maximum number of such plants per unit area, e.g. per 10 square metres. Because of the difficulty of recognizing cultivar impurities, the crop inspection has to be supplemented by the examination of a control plot.

Limits may also be set on the incidence in the crop of:

plants of certain other species, e.g. another cereal in a cereal crop
noxious weed plants
plants infected with specified seed-borne disease, such as the smut and bunt diseases of cereals
plants of the seed-bearing parent shedding pollen in the production of an F1 hybrid cultivar

These limits may be expressed in terms of percentages or numbers per unit area.

The seed

Conformity to the seed standards is determined by laboratory tests and carried out on a sample drawn after processing and packaging. The standards refer to some, not necessarily all, of the following attributes:

analytical purity, determined by testing a sample containing between 2500 and 3000 seeds
number of seeds of certain other crop species, such as barley contaminating wheat, counted in a sample of prescribed size at least ten times as big as the sample for analytical purity
number of seeds of certain specified noxious weeds, determined in a similarly large sample
germination capacity
health, expressed as the percentage of seeds infected with certain specified diseases
moisture content
size and uniformity of size

Conformity to the standard of cultivar purity prescribed is judged on a combination of the observations made in the field inspection and in the examination of the pre-control plot. Any obvious cultivar impurities noted in the laboratory test may also be taken into account.

It will be noted that counts of other species and of noxious weeds are made both in the field inspection and in the laboratory test. There are two reasons for this. When only an occasional contaminant is present, it may be missed in sampling the seed lot, but nevertheless observed by the inspector walking through the much bigger population of plants that make up the crop. Another reason is that these contaminants are difficult to remove in the cleaning process, and much effort is saved if certification is refused sooner rather than later. In the case of seed health, no infection may be observed in the crop, but the seed may be infected by wind-borne spores and this can be tested in a subsequent laboratory test.

Procedure

The procedure follows an annual cycle. It is initiated by a person who has a quantity of seed and submits an application to multiply it under the certification scheme. The procedure then follows in general, but with national variations, along the following lines (see figure 12.1).

Acceptance

The application is accepted only if the seed which it is proposed to sow is either basic seed or certified seed of a suitable generation, and was produced in the preceding year. The reference number of the seed lot and the quantity to be sown have to be stated. If the applicant is the breeder or maintainer of the cultivar, he will propose to multiply pre-basic seed.

The certification authority may have to be satisfied that the farm on which the seed will be sown is suitable for seed production, with adequate harvesting, threshing, and drying equipment, and that after harvest suitable processing and storage facilities will be available. One of the main criteria in judging adequacy is the capacity to deal with the peak load while still maintaining the identity of different lots. Some certification authorities accept applications only from approved processors, who have to arrange for farmers to grow the crops for them under contract. Other authorities accept applications from farmers who are permitted to sell the harvested seed to approved processors. When a grower applies for the first time, an advisory visit to the farm may be required, but applications from experienced growers can be accepted without question.

The applicant indicates the exact field in which the seed is to be sown and its area, and gives information about its previous cropping, isolation and, if it is a perennial crop, previous harvests. The field must comply in these respects with the conditions laid down by the authority.

Pre-control

A pre-control sample is required from the seed to be sown. If the seed is basic or certified, a sample will have been taken and subjected to

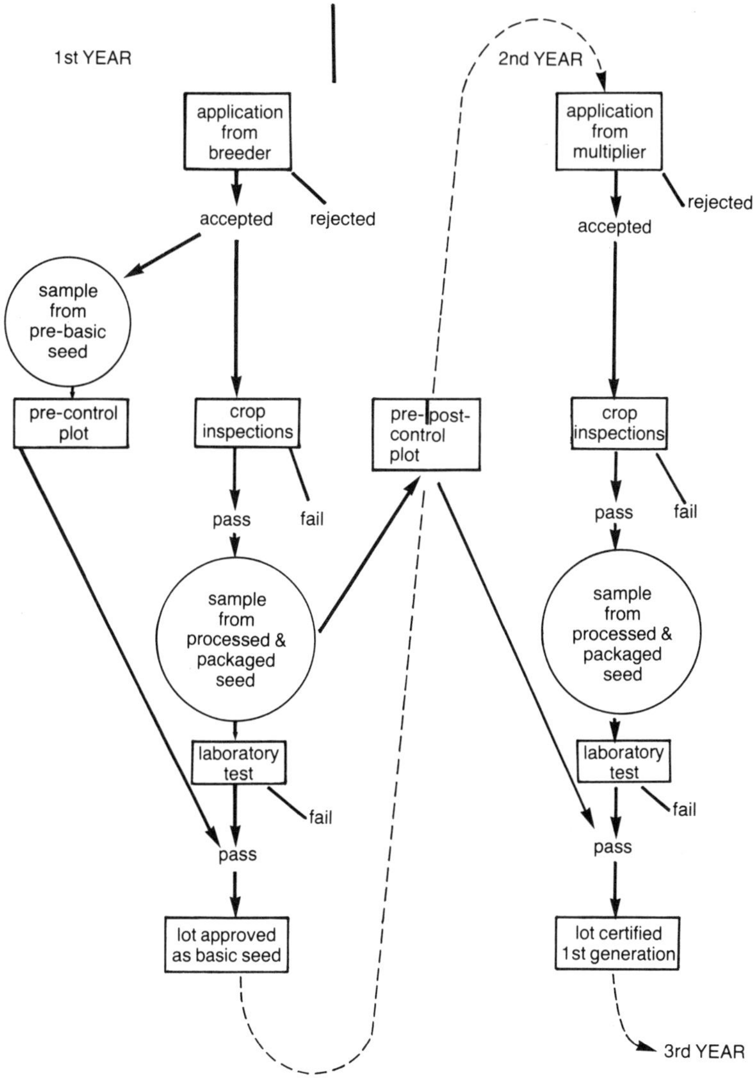

Figure 12.1 Certification procedures.

laboratory tests at the time of approval or certification in the previous year, and this sample will already be in the possession of the certification authority. If the seed is pre-basic, it is necessary to draw an official sample.

The sample is sown in a plot on the certification authority's farm. This is known as a *pre-control plot* and is used as a test for cultivar purity as described in chapter 11. One plot represents all the crops sown throughout the country with seed of the lot from which the sample was taken. It is kept under observation throughout the growing season, and the evidence is used to supplement the reports on the seed crops received from crop inspectors. The plot is sown at about the same time as the farm crops and grows parallel with them. It is an advantage, however, if the authority's farm is in an early district, so that in difficult cases observations made on the plot can be checked by visits to the farms.

Control plots sown with authentic seed are sown for each cultivar. Pre-basic seed is checked against a stock of seed maintained in store for some years by the certification authority; basic seed and certified seed are checked against pre-basic seed. For forage crops, two kinds of pre-control plots are grown—pre-basic seed as spaced plants, and basic and certified seed in rows. A detailed study of spaced plants is necessary to ensure that genetic shift has not occurred in seed being multiplied by the breeder, but the technique is too demanding in time and space to be applied to subsequent generations.

Training

Crop inspection requires a large number of specially trained staff for a short period each year. Some inspectors have other duties for the rest of the year, others are recruited temporarily, but during the inspection period all are regarded as officers of the certification authority. For this large staff, special training has to be provided in the recognition of cultivars and the techniques of crop inspection.

In any year, some inspectors have had previous experience while others are new to the work. For the latter, an introductory training course lasting about a month is necessary; this starts with plant structure in the species to be inspected, concentrating on those parts of the plant which show cultivar differences. For experienced inspectors, a short refresher course of about a week is necessary each year before they depart for the farms. Inspectors work in pairs, checking each other's observations, and an inspector in his first year is paired with an experienced man.

For training purposes, special plots are sown and the pre-control plots can also be used. It is helpful for an inspector to examine the plots corresponding to the field crops that he will be visiting, and thus acquaint himself with the impurities that are likely to be found in the crops.

In some certification schemes, employees of seed merchants and processors are allowed to carry out crop inspections, provided they have been trained and passed tests of proficiency. These inspectors are liable to have their work checked by official inspectors.

Crop inspection

Inspection of the growing crop is carried out in order to:

confirm that the crop is in fact in the field indicated in the application and check its area;
confirm that the isolation of the crop is adequate;
confirm that the cultivar is as stated in the application;
estimate the incidence of contaminants for which standards are prescribed, e.g. cultivar impurities, other species, weeds and diseased plants. How this is done is described in chapter 11.
confirm, in the production of seed of an F1 hybrid cultivar, the identity of the two parent lines, that they are planted in the correct ratio and pattern, that no more than the permitted number of plants of the seed-bearing parent are shedding pollen, and that the male parent plants are completely removed before harvest.

More than one inspection may be necessary. Cultivar differences are usually most clearly expressed at the time of flowering, but an earlier visit will reveal early rogues and off-types. In some species a late inspection can utilize such features as the grain characteristics of sorghum and pod shape in peas. For herbage crops, an inspection of the ground may be required before sowing to check its previous cropping, isolation and general suitability. For the production of F1 hybrid maize seed, three or four inspections are necessary to assess the effectiveness of de-tasselling. Biennial and perennial crops are inspected in the year of sowing, and again in the subsequent harvest years, roots and bulbs being examined as they are lifted and replanted (see chapter 6). At one of the inspections the opportunity should be taken to report on the general suitability of the farm and its equipment for seed production; such reports can be taken into consideration in accepting future applications. These visits to the farm are primarily for official purposes, but they also serve an advisory function and can help the grower to improve his methods.

In the case of maize, genetic contamination may be apparent from obvious characters in the harvested ears, such as length and diameter of the ear, number of kernel rows, colour and indentation of kernels. The crop inspection can therefore be supplemented by a post-harvest inspection of the ears.

As a check on the seed actually sown, the inspectors ask the farmer to produce the labels from the bags; the quantity of seed and the lot reference number on each label are compared with the information given in the original application.

Harvesting and processing

In due course the seed is harvested, and perhaps dried artificially on the farm. It is then transported to an approved processing plant, where it may be dried further, conditioned, cleaned, graded, treated and finally packaged. At some stage during this succession of events, it is stored in bins or bags. For complete control, all these processes should be carried out under the supervision of an inspector from the certification authority. However, such extensive supervision is impossible to provide, and in practice there is no surveillance between crop inspection and the sampling operation after the seed is packaged. During this time the seed is on trust, but farmers, merchants and processors are advised as to the procedures to be followed to prevent contamination and mixing. As an aid, temporary labels marked with a lot reference number may be issued after a successful crop inspection, and these can be attached to the containers during transport and storage to help maintain the identity of the lot. Some check is provided by the records which processors are required to keep, but there is no guarantee that the seed lot sampled and sealed by inspectors was harvested from the crop inspected by another officer some weeks or months previously. If negligence or malpractice is discovered, the sanction is withdrawal of approval in future years.

Sealing and labelling

When a lot is packaged after processing, an inspector visits the premises, draws a sample, and sends it to an official seed-testing station for test. If the tests show that the seed satisfies the prescribed standards, the bags are sealed and labelled, and the contents are thereby certified, or approved as basic seed, as the case may be.

The seal is a device, bearing the mark of the certification authority, which is attached to the mouth of the bag in such a way that access cannot be gained to the contents without either breaking the seal, or causing visible damage to the bag. This ensures that the contents of the bag are not tampered with after the sealing operation.

The essential feature of a label is the information it carries about the contents of the bag. This information may be printed on the bag or on a tag which cannot be removed without breaking the seal. This establishes the connection between the information on the label and the seed in the bag. The label carries the following information, at least:

species
cultivar
lot reference number
grade (including generation, if required)
certification authority
date of sealing
weight
treatment with fungicides or pesticides

The maximum weight of a lot is 20 tons for cereals and pulses, and

10 tons for grasses and clovers. Provided they are separately of the required standard, seed from two or more crops may, with the permission of the authority, be blended to form one lot.

The inspectors who visit processing plants to sample, seal and label seed lots, observe how the plant is managed, assessing particularly the orderliness and cleanliness which are essential in avoiding accidental contamination and admixture. Their visits have indeed an advisory as well as a supervisory purpose.

Post-control

The sample taken from a lot of certified or basic seed is big enough to sow a plot in the following year for post-control. This plot is examined in the same way as a pre-control plot. Indeed, if the lot is to be sown on farms for the production of a further generation of seed, the post-control plot also serves for pre-control.

Post-control has three purposes. It provides a check on the effectiveness of the certification scheme as a whole, by indicating the level of cultivar purity attained by the seed certified in the previous season. Secondly, it draws attention to particular cultivars which may be genetically unstable; in such a case the breeder may have to re-select his parental material. Thirdly, it indicates any seed lot which has been wrongly certified, because of accident, negligence or malpractice. By the time such a seed lot has been identified it is too late to withdraw certification. The seed has already been sown on farms, but it is possible to ensure that it is not used to produce a further generation of seed for sowing. If a number of faulty lots can be traced back to one farm or processing plant, steps can be taken to prevent its further participation in the scheme.

FURTHER READING

American Association of Official Seed Certifying Agencies (1971), *Certification Handbook*, Publication No. 23.

FAO (1975), *Cereal Seed Technology*, chapter 8, Rome.

ISTA (1971), "OECD standards, schemes and guides relating to varietal certification of seed," *Proceedings of the International Seed Testing Association*, Vol. 36, No. 3.

Mehta, Y. P. and others (1972), *Field Inspection Manual*, National Seeds Corporation, New Delhi.

Ministry of Food & Agriculture (1971), *Indian Minimum Seed Certification Standards*. New Delhi.

13

RELEASE AND REGISTRATION OF CULTIVARS

THE OBJECT OF A PLANNED PROGRAMME OF PLANT BREEDING IS TO produce new cultivars which are better in certain defined characters than existing cultivars. In the course of this programme, many potential cultivars are thrown up which are not good enough and are discarded. Eventually a cultivar is produced which the breeder considers to have sufficient merit. The characteristics of this cultivar have to be assessed in comparison with established cultivars, and a decision made as to whether or not seed should be multiplied and released for sale to farmers. This assessment involves consideration of the cultivar's

distinctness
uniformity
stability
agronomic performance
produce value

Distinctness from all other cultivars is, of course, essential because, if the new cultivar is the same as an existing one, there is no point in releasing it. Another requirement is that the specialists responsible for multiplication and certification of the seed must be able to distinguish and identify the cultivar. The distinction may be in morphological, physiological or agronomic characters, but to facilitate the maintenance of pure stocks, a morphological distinction clearly visible in the plant is best. Distinctness is not absolute, and a difference may be so fine as to have no practical value for identification purposes. It is a matter of judgment, therefore, whether the distinctness that is claimed is in fact adequate.

In assessing a cultivar's uniformity, a number of factors have to be taken into consideration. The degree of uniformity that is attainable depends on the breeding system. In a pure line, every individual plant can be identical, but within a cross-pollinated cultivar this is impossible, and there may be considerable variability. What is desirable depends on the farming system for which the cultivar is intended. For an intensive highly mechanized system in a homogeneous region, absolute uniformity within a cultivar is the ideal, but under other conditions there can be advantages in variability. The two extremes are well illustrated by F1 hybrid and synthetic cultivars of maize. The standard of uniformity that is acceptable is therefore not the same for all crop species or for all cultivars of the same species, but should be kept as high as possible. To the agronomist in charge of multiplication or certification, a non-uniform cultivar is a mixture of indeterminate purity, which cannot with certainty be identified or distinguished from other cultivars.

Stability is the capacity of a cultivar to reproduce itself over several generations without losing its distinctive characters. A stable cultivar presents no problems under controlled multiplication for certification, and the commercial farmer need renew his seed infrequently. Like uniformity, the degree of stability that is attainable is influenced by the breeding system. Pure line cultivars are stable. At the other extreme, seed of an F1 hybrid cultivar is completely unstable, but nevertheless it is possible to produce seed of the cultivar every year.

Assessment of agronomic performance is most important. A cultivar may be adequate in other respects, but unless it is attractive to the farmer, it will never be grown to any extent. Performance includes such characters as yield per hectare, maturity, strength of straw, response to fertilizers, reaction to adverse environmental conditions (e.g. drought and frost) and resistance to diseases and pests.

The value of the produce refers to the quality of what is harvested for its purpose. This covers the suitability of wheat for baking the kind of loaf that is required, of barley for brewing and of peas for canning, the length of cotton lint, the strength of jute fibres, the quality of sunflower seed oil and the digestibility of forage. These values become of increasing significance in a cash crop economy.

Organization

The breeder may belong to an official plant-breeding station or to a commercial organization, and in some countries he (and his organization) are responsible for assessing a new cultivar's merits and deciding whether or not to release it for general use. He has complete freedom in this respect, within ethical limits, but the facilities available to him for cultivar testing may be limited. The extent to which a new introduction is taken up initially is determined by the publicity it is given, but subsequently depends upon its genuine merits as appreciated by the farming community.

In some countries, tests and trials of new cultivars are organized on a national basis by an authoritative body, which may be a government agency, or a board representing breeders, seedsmen and farmers. The submission of new cultivars to such a national system may be voluntary or compulsory.

In a voluntary system, field trials and laboratory tests for assessment of performance and quality are carried out over a period of two or three years normally, and the results are published. The better cultivars may be selected for recommendation to farmers through the extension service. This arrangement leaves to the breeder the decision as to whether or not to release a cultivar for multiplication and sale. If a cultivar is recommended, it will certainly be released, but if not, in practice it will probably be withdrawn. The certification authority before accepting a new cultivar into its scheme will have to be satisfied as to its distinctive features, uniformity and stability.

Under a compulsory regime, seed of a cultivar cannot be marketed until the cultivar has been assessed and found to be satisfactory in distinctness, uniformity and stability, and in its value for cultivation and product use. The requirement for value may be that the cultivar is either at least as good as, or significantly superior to, established cultivars; this is a matter of national policy. Cultivars which are shown to be good enough are registered on an official list. This list may be supplemented by a shorter list of cultivars which are recommended to farmers. A cultivar, once registered, does not remain on the list indefinitely; it is liable to be superseded after a few years by better cultivars coming from the plant breeding stations.

Supplementary information about a cultivar's performance and its acceptability to farmers can be obtained by issuing "mini-kits" as described in chapter 2.

Tests for distinctness, uniformity and stability

Assessment of a cultivar for distinctness, uniformity and stability involves the study of individual plants sown in small plots on an experimental farm where a wide range of established cultivars is available for comparison. As the expression of cultivar characters can be modified by the environmental conditions in which the plants are grown, the farm should be typical of the region for which the new cultivar is intended. Day-length in particular can influence the expression of characters and their uniformity, and for a large country more than one testing centre should be provided. This is also a wise precaution against loss of material through drought or other unfavourable circumstances. Early ripening cultivars attract birds and protection of the plots by nets is necessary.

A prerequisite for a testing programme is an intimate knowledge of the established cultivars, the preparation of a list of characters by which cultivars can be distinguished, and a system of recording the expression of each character. For two-state characters, the appropriate alternative is expressed by a suitable symbol. Multi-state characters can be assessed by eye in accordance with a numerical key. Taking habit of growth as an example, 1 can indicate prostrate growth and 5 erect growth, with 2, 3 and 4 indicating intermediate habits. Measurable characters are recorded as dimensions or counts, e.g. length of stem, number of flowering nodes, date of flowering.

Observations and measurements are made on three kinds of plot, which provides the most useful information depending on the breeding system and the habit of growth of the cultivar.

(i) Field plot—A small plot or a few rows of plants at normal crop density.
(ii) Spaced plants—At least 100 plants grown at wide enough spacing to allow full development of each plant. This type of plot is best for cross-pollinated cultivars showing variability and when actual measurements are required (figure 13.1).
(iii) Ear-rows—Seed is threshed out of at least 100 ears and the seeds from each ear are planted in separate short rows. This type of plot is used particularly for pure-line cereal cultivars (figure 13.2).

A determination of distinctness can be made fairly critically in two years; the first year demonstrates which other cultivars the new one most closely resembles, and the second year is devoted to a detailed comparison with these cultivars. A really critical assessment of uniformity and stability may involve a great deal of work and examination of several generations. The view is usually taken, however, that if the lack of uniformity or stability is so slight that it is not obvious in two years, it is not great enough to be of significance, though difficult cases might have to be carried over to a third year. In practice, the breeder is required to submit for test samples of seed taken from two consecutive generations of breeder's seed.

For distinctness, the field plot is examined at different growth stages and compared with nearby plots of known cultivars. This gives a general impression of how the cultivar will appear as a farm crop, and distinguishing features or combinations of characters may be

Figure 13.1 Spaced plants of grasses showing contrasting habits of growth (Department of Agriculture and Fisheries for Scotland).

Figure 13.2 Ear-rows of a wheat cultivar, showing one aberrant row (Department of Agriculture and Fisheries for Scotland).

immediately obvious. This is supplemented by observations and measurements made on individual plants in either the spaced or ear-row plot. When variability is the norm, the mean value and variance are calculated for each measurable character in the cultivar under test and in a control cultivar. Whether or not the difference between means is sufficient to indicate distinctness is decided by a statistical test. The difference between two cultivars may lie in a marked difference in one character, or in minor differences in a number of characters.

In a pure-line cultivar, serious lack of uniformity is immediately obvious in a field plot, but this assessment can be made most critically by comparing the rows within an ear-row plot. In other types of cultivar, lack of uniformity is shown by excessive variability in the measurements of spaced plants.

For assessment of stability, some of the seed obtained from the breeder in the first year is held over and sown in the following year along with the second year sample. Comparison of the plots shows whether or not there is any significant difference. In a pure-line cultivar, lack of stability shows itself by variation between plants in the *same ear-row*; this is not conclusive, but indicates that the parent plant was either genetically heterozygous or out-pollinated.

Field trials

The object of field trials is to determine the characteristics of a cultivar which are of interest to the farmer.

For cereals and pulses, the most important determination to be made is of yield per hectare of grain or seed. Basically, yield is estimated by growing the cultivar in field plots of known size, harvesting and weighing the produce, and finally calculating the yield per hectare. Other field characters are assessed on the evidence of observations made on the plots by agronomists throughout the growing season.

In the case of forage and pasture crops, the material harvested, directly or indirectly through the grazing animal, is made up of stems, leaves and inflorescences, and the gross yield may be less important than the quantities produced at different times of year. A series of cuttings may therefore have to be arranged.

A cultivar does not perform identically in all situations, and trials are carried out at a number of centres in different parts of the country. The number of centres necessary for each crop species depends on the size of the country, its variability and the accuracy required, but there should be at least one trial centre in any region that is distinctive in climate, soil, altitude, latitude, or type of farming. Similarly, even at the same location a cultivar does not yield the same every year, and trials have to be repeated over several years.

Because a cultivar's yield varies from one location to another in the same year, and from year to year at the same location, the results of the trials cannot be expressed adequately as a figure of so many kilograms per hectare, but are expressed relative to a control cultivar. The control is an established cultivar, the field characteristics and performance of which are well known and relatively stable. This cultivar is included in each trial, and treated in exactly the same way as the cultivars under test. Yield can therefore be described much more meaningfully as so many percent more or less, maturity as so many days earlier or later, and straw as so many millimetres shorter or longer, than the control cultivar. Even more precise information can be obtained from the trials if two control cultivars are included.

Even at one site there can be variation. This was well shown in a classical experiment carried out at the Rothamsted Experimental Station in England. In a field of wheat ready for harvest and which to the eye appeared to be uniform, an area of one acre (about 0.4 hectare) was marked off. This area was sub-divided into 500 plots of about eight square metres; each plot was harvested separately and the produce weighed. The average yield per plot was 1.8 kg, but actual yields ranged from 1.2 to 2.3 kg, i.e. up to 30 percent more or less than the average. One particular plot in this experiment yielded 1.4 kg and the adjacent plot yielded 2.3 kg, i.e. about 60 percent more than its neighbour. It is clear that if an attempt had been made to estimate the yield of this cultivar on the basis of one plot in this field, the result could have been very far wrong. This variation is due to soil differences (fertility, texture, moisture availability, etc.) and to the uneven distribution of pests and diseases, supplemented by accidental differences in seed rate and fertilizer distribution.

This variation is inevitable, and the techniques of field experimentation take it into account, notably by growing a number of plots of each cultivar (known as replicate plots) scattered over the experimental area. The greater the number and size of the plots, the nearer is their average yield to the true yield, but beyond a certain number and size there is little advantage to be gained. In practice there are usually four to six replicates and the plots are up to 25 m^2 in size. Replicating the plots not only improves the accuracy of the average yield, but the differences between the plots provide a measure of the variation, and this can be taken into account in drawing conclusions.

The magnitude of the variation within a trial, however, is much less than that between trials due to differences in location or season. Trials at different sites in the same year are subject to differences, e.g. in soil, latitude and weather; trials at the same site in different years to differences in weather and incidence of disease. This variability is well illustrated by figures from trials of winter wheat cultivars held in Britain in 1969 and 1970 (Table 13.1). In each year trials were carried out at six centres. One of the new cultivars under trial was Maris Templar and a control cultivar included in all trials was Cappelle-Desprez. The overall average yield of Maris Templar was 5675 kg/ha, but in 1969 it varied at different centres from 3931 to 6486 and in

Table 13.1 Yields of a new cultivar and its control in trials at six locations in Britain in 1969 and 1970.

	Yield of new cultivar (Maris Templar)				*Yield of control cultivar (Cappelle-Desprez)*	
	1969		1970		1969	1970
	kg/ha	As % of control	kg/ha	As % of control	kg/ha	kg/ha
Location 1	3931	109	6963	131	3604	5331
Location 2	6486	116	7296	121	5584	6019
Location 3	4408	100	5228	110	4408	4757
Location 4	5529	114	6229	120	4850	5209
Location 5	5003	110	6591	113	4544	5852
Location 6	4950	99	5478	115	5000	4761
Average	5051		6298		4665	5322
Average of all trials	5675 kg/ha				4994 kg/ha	

1970 from 5228 to as high as 7296 kg/ha. Comparing the two years, yields of Maris Templar in 1970 were higher at all centres to the extent of 25 percent on the average. With so much variation between centres and between years, it would be difficult to assess this cultivar's value to the farming community on this evidence alone, but comparison showed that it yielded at least as well as the control cultivar in every trial except one (at centre 6 in 1969) and on that occasion it was very close. As Cappelle-Desprez was a reliable and widely grown cultivar, it could therefore be concluded on this limited number of trials that Maris Templar was a higher-yielding cultivar.

What is the degree of superiority that a new cultivar needs to have over established ones for acceptance? This is a matter for judgment in the light of local circumstances. What is the value of the improvement to the farmer in terms of cash or of labour? Is the improvement restricted to one crop character or is it a general improvement? What is the capacity of the farmer to exploit the increased potential? How does the improvement compare with the variation between farms due to other factors? In herbage crops, differences in yield of forage due to management are in general greater than differences between cultivars. In Britain the overall losses in arable crop products in harvesting and storage are estimated to be about 10 percent of their total value, so that theoretically production could be increased to this extent by better harvest and storage techniques. In tropical countries, grain losses in storage alone are much higher. There is no universal answer to this question, but in highly developed farming systems it is usually assumed that a new cultivar is not worth releasing unless it shows a yield superiority of at least 5 percent.

The relative importance of location and season depends on the size and diversity of the country, and on the vicissitudes of climate from year to year. How many trials are necessary to demonstrate this degree of superiority can only be learned by experience. As an example, the number required in Great Britain, which is not large but fairly variable geographically and between seasons, is about 60 spread over three years.

Laboratory tests

Laboratory tests are used to assess the quality of the produce. The material for these is taken from the produce harvested at each trial and sent to a central laboratory for tests.

For assessment of disease susceptibility, field observations can be supplemented by tests in growth chambers or glasshouses.

FURTHER READING

FAO (1975), *Cereal Seed Technology*, Chapter 1, Rome.

Silvey, V. & Fiddian, W. E. H. (1972), "The interpretation of genotype X environment interaction in cereal variety assessment," *Journal National Institute Agricultural Botany*, Vol. 12, p. 477.

14

TOLERANCES AND SAMPLING

Variation in test results

When several laboratory tests are carried out on samples from the same lot of seed, the results are not all the same. This variability is a feature of laboratory testing which has to be recognized and taken into account in interpreting the results. It is due to sampling variation, experimental error, interpretational variation and lapse of time.

Sampling variation

This variation arises because the sample used in a test is very small, and the lot from which it is drawn is a mixture—of pure seed and impurities, of live seeds and dead seeds.

In a purity test, every seed and every particle has to be examined by the analyst; in a germination test, each seed may take 1 cm^2 of space in the controlled environment of a germinator. Seed testing is therefore expensive in terms of analysts' time, equipment and materials, and the work is not evenly spread over the year. So seed-testing stations have tended to reduce samples to the smallest size that will ensure an acceptable degree of accuracy in the majority of tests. The general rule is that the bigger the sample, the greater the accuracy, but beyond a certain sample size the increase in accuracy is not sufficient to justify the extra work and expense involved.

A 20-ton lot of wheat contains about 400 million seeds. Of these, about 30 000 are sent to the laboratory, but only about 3000 are used in a purity test, and a mere 400 in a germination test. If the true germination capacity of the lot is 80 percent, the lot will contain about 320 million live seeds and 80 million dead ones. From so many it is theoretically possible to take 400 live seeds, or even 400 dead seeds, for a germination test, giving test results of 100 percent or nil. Though possible, this is highly improbable and, with proper sampling, if a number of tests are performed, most of them will produce results in the region of 80 percent.

Sampling variation can be demonstrated by taking a quantity of seed, such as white mustard, dyeing one-fifth of the seeds red or some other colour, and drawing samples of 100 seeds with a laboratory counter as though for a germination test. If the white seeds are regarded as alive and capable of normal germination, and the red seeds as dead, the nominal germination capacity of the seed is 80 percent. In an actual exercise of this kind, the numbers of white seeds in 16 draws of 100 seeds were:

77, 83, 78, 83 73, 74, 81, 89 79, 85, 80, 75 78, 82, 78, 80

The average is 79.69, which is near enough to the correct proportion of 80 percent, but the actual figures range from 73 to 89 percent.

If these replicates of 100 seeds are combined into successive fours (simulating a germination test of 400 seeds), the percentages of white seeds become, in whole numbers:

80, 79, 80, 80

This illustrates the fact that if the sample size is increased, the variation is decreased. Statistical calculations show that if the true germination capacity is 80 percent and 1000 tests of 100 seeds each are carried out, 990 of them will give results between 71 and 89, i.e. within a range of 18. If, however, the sample size is increased to 400 seeds, the corresponding range is reduced to 9 (76 to 85 percent). By increasing the sample size four times, therefore, the variation is reduced by half. The actual variation due to sampling depends on the germination capacity; theoretically it is at a minimum at 100 percent germination, and reaches its maximum at 50 percent germination (see figure 14.1). Sampling variation in purity tests is less than in germination tests because the number of seeds in the sample (2500 to 3000) is much greater.

Experimental error

Differences in test results can arise from faults in the laboratory procedure or in the apparatus.

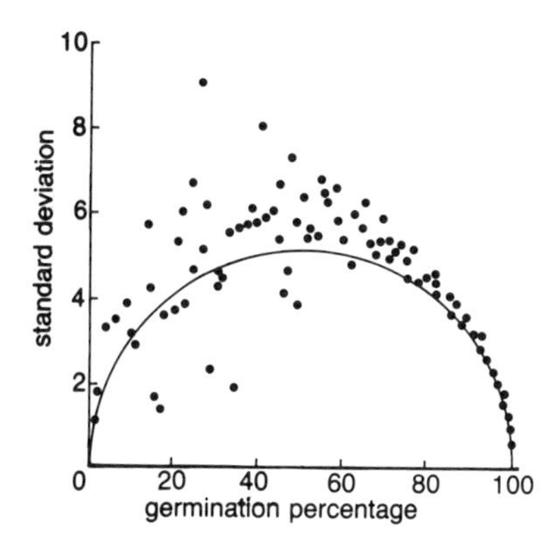

Figure 14.1 Variability in laboratory germination tests. In a test the germination capacity is taken as the average of the number of seeds germinating in 4 replicates of 100 seeds. The continuous line shows the standard deviation to be expected theoretically within tests at germination capacities ranging from 0 to 100. The black dots indicate the average of the actual standard deviations found in 20 000 routine tests of different species (redrawn from *Journal NIAB*, Vol. 9, 1963).

In a purity test, if the air is hot and dry, and the analyst works slowly, the moisture content of the seeds and the weight of some parts of the sample may change during the test. The sampling device may be faulty, the balance may be inaccurate, or the analyst might identify a seed wrongly. Accidents and mistakes can happen occasionally, but in well-managed laboratories, variation due to experimental error in purity testing is on the whole insignificant.

In determining the result of a germination test, two important factors are temperature and moisture supply, and these are not always under complete control. These factors vary for climatic reasons and because of difference in the equipment used. Good seed is capable of germinating vigorously in different environments, and the exact conditions of a test are not very important. This, however, is not true of poorer-quality seed, which has been harvested in unfavourable weather, stored for a long time, or subjected to poor storage conditions. Such seeds are more sensitive to test conditions, and slight differences can influence the result. This is illustrated in figure 14.1 where the semicircular curve shows the variation to be expected between sub-samples in the same test, due to sampling only, while the dots show the variation actually found. From 100 down to about 80 percent germination, the agreement between the theoretical sampling variation and the actual variation is good, proving that the experimental variation is negligible. At lower germinations, however, the variation can be far in excess of the sampling variation. Experience shows that this variability is greater between tests carried out in different laboratories than between tests within the same laboratory.

Interpretational variation

As explained in chapter 1, definitions form an important element of the rules for seed testing. Because different definitions were used in different countries, it became necessary to negotiate definitions which every seed-testing station was prepared to accept. However, that is not enough, for analysts do not always understand and interpret the definitions in the same way, and these interpretational differences can give rise to serious discrepancies.

The International Seed Testing Association has striven to minimize interpretational differences in three ways:

by modifying the rules from time to time in the light of experience
by organizing international training workshops for analysts
by distributing samples from the same seed lot to all member laboratories for testing and then comparing the results.

Differences have not yet been completely eliminated, but they are now much less than in the past. This is the aspect of seed testing that demands more skill and experience from the analyst than any other.

In purity tests, difficulties arise in interpreting broken seeds and grass seeds with poorly developed caryopses, but interpretational variation is no longer a significant cause of difference.

In a germination test, a seed is not regarded as having germinated unless it produces a normal seedling. Whether a seedling is normal or abnormal has to be judged by the analyst, and a seedling which is regarded as normal by one analyst may be regarded as abnormal by another. The International Seed Testing Association has dealt with this difficulty by providing precise descriptions of normal and abnormal seedlings of many species. Nevertheless, interpretational differences of this kind are still significant; they are at a minimum within the same laboratory, or within a group of laboratories under unified technical control, but can be serious between tests carried out in different countries. Abnormal seedlings appear most frequently in tests of poor-quality seed, and this is an additional explanation of the high variability that occurs in germination tests of such seed.

Lapse of time

A test result refers only to the seed lot at the time of sampling and testing. If some time elapses between two tests, the condition of the seed may have changed. Damage by rodents or insects may give rise to a lower-purity percentage. Germination capacity may have fallen due to poor storage, or may even have been enhanced by the disappearance of dormancy.

Tolerances

If variation in test results is inevitable, how much difference between tests is acceptable? This question arises in relation to process control, commercial transactions and law enforcement.

To take as an example a seed lot with a germination capacity of 80 percent, if a thousand tests each of 400 seeds are carried out, a wide range of results is to be expected. The number of tests in which any particular germination percentage occurs, allowing for sampling variation only, can be calculated mathematically, and the frequency of differences from a true germination percentage of 80% are shown in Table 14.1 as an example. A difference of, say, 2 percent occurs so often that it is of no significance and can be ignored. A difference of 9 percent is so unusual as to cause suspicion that it might be a real difference in germination capacity.

Table 14.1 Seed of 80% germination—Number of tests per 1000 expected to give results differing from 80% to the extent indicated in column 1.

Difference	*Number of tests*
4% or less	950
5% or more	50
6% or more	25
7% or more	10
9% or more	1

Similar calculations can be made for each percentage of germination. The limit of acceptibility for each percentage is called a *tolerance* and lies somewhere between the extremes noted in the last paragraph. A commonly used tolerance is the least difference to be expected in 50 tests out of a thousand—or, put in another way, the greatest difference to be expected in 950 tests out of a thousand. In the case of 80 percent germination capacity, this would be 5 percent for standard tests of 400 seeds (Table 14.1). However, in certain circumstances a wider tolerance may be applied. In the international rules, the tolerances for purity and germination tests are based, not on theoretical calculations of sampling differences, but on the actual variation found in tests carried out in different laboratories. Tables of tolerances which can be used in different circumstances are to be found in ISTA's *Handbook of Tolerances*.

Sampling

A laboratory test measures the purity, germination capacity, or other attributes of the *sample*. The result can be applied to the seed lot as a whole only if the sample is representative of the lot, i.e. if pure seed and impurities, live seeds and dead seeds, etc., are present in the sample in the same proportions as in the bulk.

If the components of a seed lot were uniformly distributed throughout, it would be sufficient to take a handful from one point in the lot and use it as a test sample. In practice, however, a seed lot is never completely uniform and, if handfuls are taken at different points, the components occur in different proportions. The reasons why seed lots are never quite uniform can be summarized as follows:

1. Segregation by gravity of heavy and light seeds within the bulk or within a bag.
2. Differences within the crop from which the seed was harvested. At different places within the field there may be variation in maturity, lodging, disease or the occurrences of weeds.
3. Duration of harvesting operations. If harvesting the crop is interrupted by bad weather, the condition of the seed will not be the same before and after the interruption.
4. Lack of uniformity in threshing and subsequent processing and storage of seed from the same crop, e.g. different machines, differences in the adjustment of the same machine, different storage conditions.
5. Putting the seed from two or more crops together to form one lot.
6. Failure to blend the lot adequately before packaging.

It should be appreciated, however, that even thorough blending does not ensure perfect uniformity. Perfection would imply an orderly arrangement of weed seeds, etc., throughout the bulk, and the best that can be achieved in practice is a random distribution. Some variation is therefore inevitable, and to obtain a representative sample it is necessary to take seed from numerous points throughout the lot.

A lot may *appear* to be uniform because all the visible impurities, such as broken seeds and weed seeds, have been removed in the cleaning process. Nevertheless, seeds of a different cultivar, dead seeds and diseased seeds, though not apparent to the eye, may be irregularly distributed.

To obtain a sample, numerous small quantities, called *primary samples*, are taken at random from different points in the lot A (see figure 14.2), and mixed thoroughly to form a *composite sample* B. Small quantities are then taken from B and mixed to form a *submitted sample* C. This process is repeated until a sample is obtained of the size required for testing E; this is known as the *working sample*. A modification that may be more convenient in the second and subsequent stages is to reduce the sample size by half at each stage, instead of taking out a number of small quantities (see page 189).

Because it is more difficult to draw a small representative sample

Table 14.2 Examples of lot and sample sizes prescribed in the International Rules for Seed Testing.

	Maximum weight of lot	*Minimum sample weights*	
		Submitted sample	Working sample for purity tests
	kg	g	g
Agrostis spp.	10 000	25	0.5
Medicago sativa	10 000	50	5
Pennisetum typhoides	10 000	150	15
Cannabis sativa	10 000	600	60
Sorghum bicolor	10 000	900	90
Triticum aestivum	20 000	1000	120
Dolichos lablab	20 000	1000	500
Zea mays	20 000	1000	900
Vicia faba	20 000	1000	1000

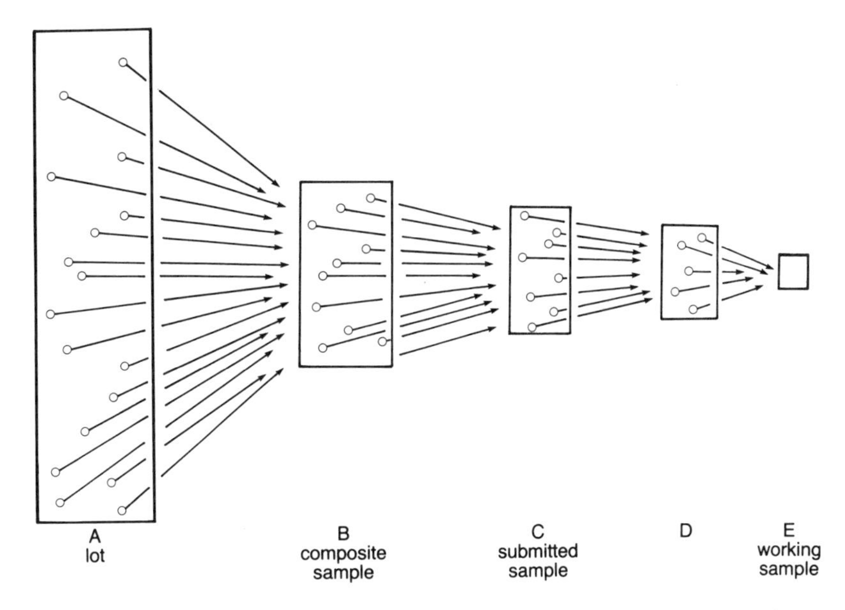

Figure 14.2 Diagrammatic representation of successive stages in taking a sample from a seed lot.

than a large one, sampling is done in two stages. In the first stage a large sample (the submitted sample) is taken in the warehouse and sent to the seed-testing station. This sample is usually about ten times bigger than the quantity required for testing, and the final stages of drawing the working sample are performed in the laboratory, where more refined apparatus is available.

Warehouse sampling

The size of a lot that can be represented by one test sample depends really on the uniformity of the lot, but this is an unknown variable. ISTA, therefore, sets the following limits based on seed size, and these are generally followed by certification and law enforcement authorities (see Table 14.2):

Seeds as large as wheat or larger—not more than 20 000 kg
Seeds smaller than wheat—not more than 10 000 kg

The number of primary samples to be taken depends on the size of the lot. The following are the minimum numbers:

Lot size	*Number of primary samples*
Up to 50 kg	3
51–500 kg	5
501–3000 kg	One for each 300 kg, but not less than 5
3001–20 000 kg	One for each 500 kg, but not less than 10

If the lot is in bulk, the primary samples are taken at different places and at different depths.

If the lot is in bags, the number of primary samples is in accordance with the weight of the lot as shown above. Except in very small lots, not more than one sample is taken from any one bag, but the actual position in each bag should vary. The bags to be sampled are picked at random from the whole lot, and are not the most easily accessible ones.

Bags are usually sampled with an instrument which leaves a hole in the material. If the bag is of cloth, the threads are not broken and can be returned to their place. If the bags are of paper or plastic, the hole can be repaired with an adhesive patch.

The primary samples are thoroughly mixed in a suitable container to form the composite sample. This sample should be at least four times more than the quantity required for the submitted sample and can be reduced by means of apparatus similar to that used in the laboratory, but of larger size. Alternatively, if it is difficult to mix and reduce the sample properly under warehouse conditions, it may be possible to arrange for the entire composite sample to be sent to the laboratory for reduction there.

Sample weights of a few species are given as examples in Table 14.2. They depend on the size of the seed, but with minimum and maximum submitted samples of 25 and 1000 g respectively.

The submitted sample is sent to the laboratory in a small bag, firmly closed and strong enough to withstand whatever form of transport is used. Information sufficient to identify the lot with its sample is enclosed, such as species, cultivar, reference number, size of lot, date and place of sampling and name of sampler.

If the sample is to be used for the determination of moisture content, the whole operation should be carried out quickly, with minimum exposure to the atmosphere, lest the moisture content change. The primary samples should be drawn into a polythene bag and the submitted sample enclosed in a sealed moisture-proof container.

Various devices are available for abstracting primary samples from bags, from bulk or from a packaging machine. They may not, however, be suitable for chaffy seeds which do not flow freely. For seed of this kind, it is permissible to take the sample by hand. The open hand is inserted into the bag, or bulk, the hand is then closed and held tightly closed while the handful of seed is withdrawn.

The stick

This consists of a double metal tube with a pointed end (figure 14.3). Both tubes have coinciding slots, which can be opened or closed by turning the inner tube within the outer one, and the inner tube is divided into sections by transverse partitions. In use, the stick is

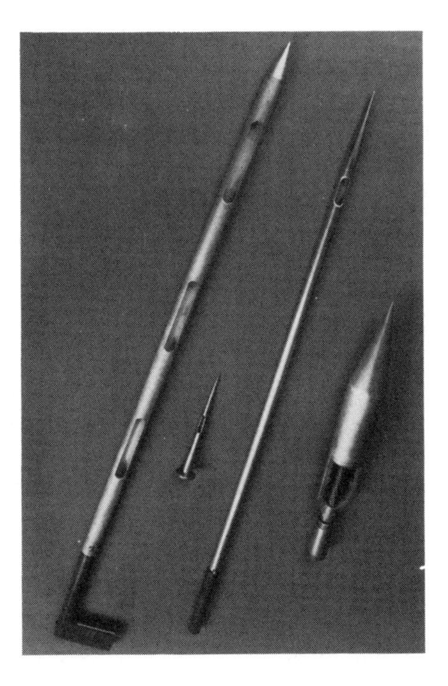

Figure 14.3 Instruments for drawing primary samples from a seed lot. From left to right: stick, thief, spear and bulk sampler (head only with long rod removed) (Department of Agriculture and Fisheries for Scotland).

inserted into the seed mass with the slots closed; when the slots are opened, seed flows into the inner tube, and this can be facilitated by one or two opening and shutting movements. The slots are finally closed, the stick withdrawn, and the contents (which constitute a primary sample) transferred to a suitable container.

Sticks are made in various diameters to suit different species of seed, depending partly on the size of the seed and partly on the ease with which they flow. They are also made in different lengths, about 750 mm for sampling bags, and three or four metres for sampling bulks.

In sampling a bag, the stick is inserted diagonally from top to bottom (figure 14.4). A bulk may be contained in a bin or a box, and the depth that can be effectively sampled is limited by the length of the stick. The stick is inserted diagonally downwards at a number of points, depending on the number of primary samples required.

The spear

The spear is a metal tube with a sharp solid point and an oval hole just behind the point (figure 14.3). It is suitable only for seed in bags.

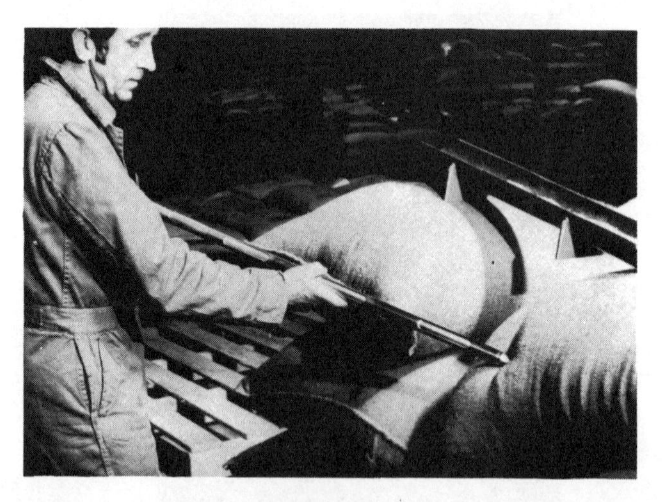

Figure 14.4 Drawing a primary sample from a bag with a stick sampler (Scottish Agricultural Industries).

In use, the spear is thrust upwards into the sack at an angle of 30 degrees and with the hole facing downwards, until the point reaches the opposite side. It is then turned until the hole faces upwards, and this allows seed to flow down the spear into a small bag or container held by the sampler. The spear is withdrawn at a steady speed so that equal quantities of seed are withdrawn at all points along its line of movement. If the seed flows with difficulty, the spear can be gently agitated to maintain a uniform flow. The seed withdrawn at each insertion constitutes a primary sample. Normally a bag need not be sampled at more than one point, but throughout the lot primary samples should be taken at different levels, e.g. top, middle and bottom.

Spears, like sticks, are made in various diameters to suit different species of seed, and the length is sufficient to penetrate to the farther side of the bag. The short spear, called a "thief" (figure 14.3), draws seed only from the outermost parts of the bag and should not be used.

Bulk samplers

Bulk samplers are devices for drawing samples from very deep bulks. One type is a large version of the stick sampler. Another (figure 14.3) consists of a heavy metal cup with a close-fitting lid, attached to the end of a metal rod to which additional lengths can be added according to the depth to be sampled. The cup is pushed vertically downwards into the bulk, the pushing action keeping the lid pressed to the cup. When the rod is pulled to withdraw the cup, the lid is

Figure 14.5 Conical divider used for drawing laboratory samples (Department of Agriculture and Fisheries for Scotland).

opened sufficiently for the cup to fill instantly with seed. Each cupful constitutes a primary sample, but only from the point at which the cup filled, and not from the line along which it was withdrawn.

Mechanical samplers

Primary samples can be drawn automatically from seed as it passes along a conveyor or down a chute towards a packaging machine. It is important that the instrument should take seed uniformly from the entire cross-section of the seed stream and that seeds do not bounce out of it.

Laboratory sampling

From the sample that arrives from the warehouse, a working sample about one tenth of its weight has to be extracted for a purity test. The methods commonly used follow one or other of two principles—

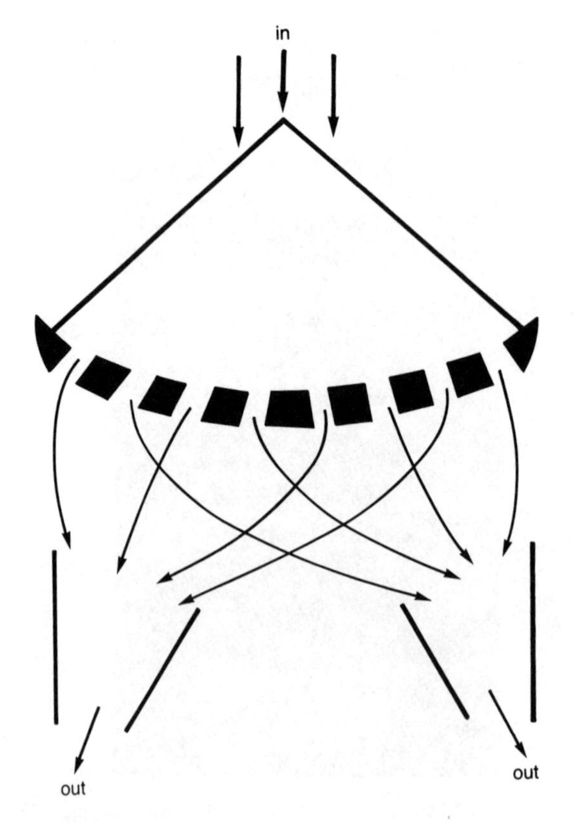

Figure 14.6 Conical divider used for drawing laboratory samples.

successive halving, or extraction and combination of small quantities—combined with thorough mixing. The working sample need not be of the exact weight required but, of course, it is weighed accurately before the analysis begins.

Successive halving

In this method, the submitted sample is divided into two approximately equal portions; one portion is discarded and the other halved again. This halving procedure is repeated until a sample of approximately the quantity required is obtained. Before each halving, the material is thoroughly mixed. Various types of apparatus are available; the most commonly used are the conical, the multiple-slot and the centrifugal dividers.

The conical divider (figure 14.5) consists of an inverted metal cone with numerous ports closely spaced around its circumference (figure 14.6). Alternate ports lead through ducts to a common outlet on one side, and the others lead similarly to an outlet on the opposite side. When the seed is poured over the cone, it flows through the ports, half of it emerging through the left outlet and half through the right outlet.

In the multiple-slot divider, rectangular ports are held in a frame, with alternate ports leading to left and right. The seed is poured evenly over the frame and flows down through the ports, half to one side and half to the other (figure 14.7).

In the centrigual divider, the seed falls on to a rapidly rotating disc, which throws half of it out to one side and half to the other (figure 14.8).

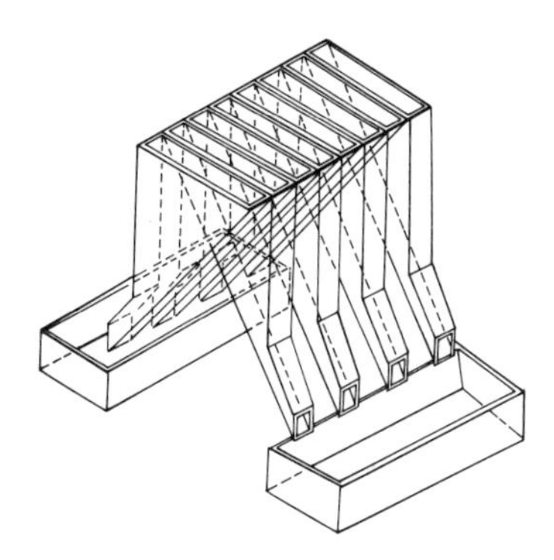

Figure 14.7 Multi-slot divider used for drawing laboratory samples (redrawn from International Standards Organization).

Figure 14.8 Centrifugal divider for drawing laboratory samples (Scottish Agricultural Industries).

Extraction and combination of small quantities

This principle is best illustrated by the random cups method; this is particularly suitable for small seeds, which do not bounce or roll, with a working sample up to about 10 g, e.g. grass and clover. Six to eight small cups are placed at random on a tray; the sample is poured evenly over the tray, and the seed that falls into the cups is taken as the working sample (figure 14.9).

The size of the cups depends on the sample size required, and on the size of the seeds, but a number of sets ranging from 10 to 20 mm

Figure 14.9 The random-cups method of drawing laboratory samples (Scottish Agricultural Industries).

in diameter and depth covers most requirements. The cups are usually made of brass, but other material may be used, provided it does not develop static electricity and the cups are heavy enough to be stable.

Sampling for a germination test

For a germination test, four replicates of 100 seeds are taken from the pure seed fraction of a purity test. For this purpose a vacuum seed counter, or planter, is frequently used, which not only retains 100 seeds, but also spaces them ready for planting (figure 16.4). In the circular head there are 100 holes, one seed is held to each hole by suction from an air pump and, when the suction is cut off, the seeds drop into place on the medium. Different sizes are available for different species of seed.

FURTHER READING

ISTA (1976), "International Rules for Seed Testing," chapter 2 and annexe 2, *Seed Science & Technology*, Vol. 4, No. 1.

"Handbook of tolerances and measures of precision for seed testing," *Proceedings of the International Seed Testing Association*, Vol. 28, No. 3, 1963.

15

TESTING FOR PURITY

WHAT IS MEANT BY ANALYTICAL PURITY, SPECIES PURITY AND FREEDOM from weeds has been outlined in chapter 1. In this chapter more details are given of the testing procedures in the laboratory, but not to the extent that is necessary for training analysts.

Analytical purity

In this test a working sample is separated into three fractions:

pure seed
seed of other species
inert matter

The sample is of a weight which on average contains 2500 to 3000 seeds; some examples of sample sizes for different species are given in Table 14.2. The normal procedure is for a sub-sample of this size to be taken from the sample submitted to the laboratory (by one of the methods described in chapter 14) and transferred to a dark smooth surface. Every seed or particle is then examined by the analyst, identified and put into one of these fractions. When the separation is complete, each fraction is weighed, and its weight expressed as a percentage of the total weight. Analytical purity is the percentage of pure seed determined in this way.

Pure seed

Pure seed is seed of the species that the sample purports to be, no attempt being made to distinguish between cultivars within the species. This seems simple and obvious enough, but there are three difficulties—identification, the nature of the seed, and the categorization of immature and broken seeds.

The art of identifying seeds can be learned only be experience—it needs a sharp eye, trained to recognize minute differences. An essential item of laboratory equipment is a collection of named seeds belonging to a wide range of crop and weed species. The common species become familiar, but identification of the rarer impurities may require comparison with authentic samples.

In some cases it is impossible to distinguish with certainty between seeds of two species belonging to the same genus. For example, two species of ryegrass, *Lolium perenne* and *L. multiflorum*, can be distinguished only by the presence of awns on the seeds of the latter, but the awns are liable to be broken off in the cleaning process, and it is then impossible to say to which species the seed belongs. In such a case, the pure seed can be identified and described as belonging to the genus, but not to any particular species.

That the sample as a whole does indeed belong to the species claimed for it, can sometimes be verified by special tests. A chemical test may be available, or seeds can be germinated and the seedlings grown until they show characters by which the species can be identified, e.g. in the genus *Lolium*, the two species mentioned above can be distinguished by a fluorescent substance which diffuses out of the young roots of the seedlings. Such tests, however, are not regarded as part of a purity test, partly because of the time required and partly because they are not precise enough.

As explained in chapter 3, the seed that is sown by the farmer is not necessarily a seed in the true botanical sense. It may be a fruit or part of a fruit containing one (or, rarely more) true seed, and may be surrounded and hidden by additional appendages. For testing purposes, pure seed is normally regarded as seed in the farmer's sense.

A whole seed is regarded as pure seed even if it is undersized, immature, diseased, or has germinated in store. In the case of some cereal and most grass species, the caryopsis is hidden within the chaffy floret, but the analyst has to confirm that it is present; even if its development is only partial, the floret is regarded as pure seed. A broken seed is, with some exceptions, regarded as pure seed if the piece is judged to be more than half the original unbroken seed. Empty florets and broken pieces less than half size are treated as inert matter.

The essential point about undeveloped and broken seeds is that the analyst does not pass judgment as to whether or not the seed is germinable. Empty grass florets (and broken pieces half-size or less) are assumed to have no planting value; even if the latter were viable, they would not give rise to robust plants in the field. Partially developed seeds (and broken seeds more than half-size) may or may not be viable; this and the value of the seedlings they produce is judged in the germination test. If they are non-viable, they add to the purity percentage, but detract from the germination percentage; that is why the results of the purity and germination tests must always be considered together when evaluating a seed lot.

Other seeds

Into this fraction are put seeds of any species other than the pure seed species, and each seed has to be identified. This fraction includes both crop seeds and weed seeds, and the percentage of any species is reported if it is greater than 0.1. Empty florets (and broken seeds less than half-size) are treated as inert matter, as in the case of the pure seed species.

Inert matter

This fraction takes everything in the sample that is neither "pure seed" nor "other seeds". It includes seed material such as empty grass florets and broken seed pieces half-size or less; other plant material such as chaff, awns, leaves, and pieces of stem; mineral matter such as soil, sand, and stones. For convenience it also includes fungus bodies such as ergot, and insect galls, even though (strictly speaking) these are not "inert".

Separation

Identification of each seed in the sample is by visual characters. Normally the examination is by the unaided eye, supplemented by a hand lens or binocular microscope for doubtful seeds. This is entirely adequate for large seeds such as pulses and cereals; but analysis of small forage seeds requires acute eyesight and may cause strain. A recent development is the invention of a machine which moves the seeds in single file under a binocular microscope; the analyst has both hands free to stop the flow and subject any doubtful seed to more prolonged scrutiny.

In many grass species the seed is a chaffy floret, and it is not sufficient to identify the species; the analyst must also confirm that each floret contains a caryopsis. This can be done in either of two ways—by touch or by transmitted light. By feeling with a finger tip or with a spatula, an experienced analyst can tell whether or not a floret is empty. If the florets are placed on a sheet of ground glass with a light underneath, full seeds can be distinguished by the dark shape of the caryopsis showing through the floret.

Figure 15.1 Blower used for separating light from heavy seeds in the uniform blowing method of purity analysis (Department of Agriculture and Fisheries for Scotland).

The task of analysing grass seed can be lightened by using a piece of apparatus which blows an adjustable current of air through the working sample and separates light from heavy seed. Using this blower, an analyst can separate the sample into three parts containing full seeds, empty seeds, and doubtful seeds. A quick examination confirms that the full and empty seeds are indeed so, and a close examination of individual seeds is necessary for the doubtful seeds only.

A refinement is the uniform blowing method in which the sample is blown for three minutes at a pressure which will give a sharp separation of empty and full seeds, leaving no "doubtfuls". The blower used in this method (figure 15.1) is capable of producing a uniform current of air at a predetermined pressure. A sharp separation is not possible in many species, and the method is applicable so far only to *Poa pratensis, Chloris gayana* and *Dactylis glomerata*, each species requiring a different pressure. After separation, obvious impurities such as seeds of other species, pieces of stem, sand, etc., have to be removed by the analyst. This method serves not only to reduce the analysts' work load, but also to promote uniformity in test results. There is no difficulty in categorizing seeds that are full or empty, but analysts may differ in their treatment of the doubtful seeds which lie between. The uniform blowing method separates these doubtful seeds mechanically into the full or empty category according to their behaviour in a standard current of air, and the analyst does not have to make the decision. The separation may not be absolutely correct, but serious errors are rare. The important point is that, when laboratories in different countries use the same standardized blower, they all obtain the same result in their tests of the same seed.

A similar situation exists as to the use of sieves. These can be used in the analysis of some species to make a preliminary separation between pure seed and other seeds or inert matter, and the separation is then checked by visual examination. In the case of beet, however, it is permissible to make a positive separation between pure seed and inert matter derived from beet seed, by using sieves with holes of a prescribed size.

Species purity and freedom from weeds

As explained in chapter 1, a determination is sometimes required of the *number* of seeds of other species, the other species being either cultivated species or noxious weeds. This determination requires the examination of a sample about ten times the weight of the working sample for analytical purity. This large sample is searched for seeds of the particular species in question. The test may be required for certification purposes or for law enforcement, and the result of the test is expressed as the number of seeds found in the weight of material examined, e.g. one seed of wild oat found in *x* grams of barley seed. A certification standard for weed seeds is usually in the form

No seeds of species X in one kilogram.

and in this case a sample of at least the standard weight has to be searched.

When the seed that is being sought differs in some physical characteristic (e.g. size, gravity or roughness), sieves, blowers or other mechanical devices can be used to make a preliminary separation, and so reduce the work involved.

FURTHER READING

FAO (1975), *Cereal Seed Technology*, chapter 7, Agricultural Development Paper No. 98.

ISTA (1976), *International Rules for Seed Testing*, chapter 3, "Seed Science and Technology," Vol. 4, No. 1.

Mercer, S. P. (1938), *Farm and Garden Seeds*, chapter 3, Crosby Lockwood, London.

US Department of Agriculture (1952), "Testing Agricultural and Vegetable Seed," *Agricultural Handbook* No. 30.

16

TESTING FOR GERMINATION CAPACITY AND VIGOUR

Germination testing

In a germination test, a small sample of seeds, usually 400, is given optimum conditions for germination and, after these conditions have been maintained for an appropriate time, the number of seeds that have produced normal seedlings is counted. As discussed in chapter 3, seeds have three essential requirements for germination—oxygen, water and a suitable temperature—but in certain cases other factors have to be supplied.

Oxygen

Provided ventilation in the germinator is adequate, there is normally sufficient oxygen in the atmosphere, and no special arrangements are necessary.

Water

Seeds do not germinate well when immersed in liquid water. The reasons for this are not fully understood. One reason is that an immersed seed cannot obtain sufficient oxygen for respiration; another may be that soluble substances are leached out of the seed. In germination tests, water is supplied by absorption through a *medium* or *substratum*, such as absorbent paper or sand. The essential feature of a medium is that it holds a fair quantity of water, but gives it up freely to the seeds. At the same time it includes numerous cellular spaces, which are not filled with liquid, and which allow access of air to the seeds. The material should not of itself provide a suitable medium for the growth of saprophytic micro-organisms—otherwise, these would spread more easily from one seed to another. Another requirement is that the material should be free from toxic substances.

The materials most commonly used are absorbent paper and sand. Small pieces of paper are usually called *blotters* and large sheets are in the form of *towels*. Small seeds are placed on top of a blotter or between two blotters; large seeds are placed between layers of towels, which are rolled (figure 16.1) and kept in a vertical position (figure 16.2). Sand is suitable for large seeds only, because small ungerminated seeds are difficult to find in it at the end of a test. The seeds are pressed into the sand and buried. Paper can be used once only, but sand can be used repeatedly, provided it is sterilized between tests.

The medium must remain moist throughout the test, but it should not contain so much water that the air spaces are filled. Suitable paper is capable of holding enough water for the duration of the test, provided sufficient thicknesses are used and the atmosphere of the germinator is kept at a relative humidity of more than 90 percent. Alternatively, a blotter can be kept moist by an absorbent wick which draws up water from a tank in the germinator. Tests on sand are carried out in boxes or dishes with tight-fitting lids to prevent drying out. It is not advisable to add water during a test, partly because it may not be distributed uniformly to all the seeds, and partly because there is a tendency to add an excessive amount.

Temperature

The temperature at which the seeds are maintained is that which is expected to give the quickest and most complete germination of the sample. This depends partly on the species and partly on the stage of after-ripening, the optimum temperature for non-dormant seeds being usually higher than for dormant seeds. The most suitable temperatures have been determined by experience and experimentation. For a particular species, one temperature may be the best for all samples; for others there may be alternatives, and the analyst selects the one which he has found to suit the majority of samples tested in his laboratory. This depends mainly on the degree of dormancy normally present.

Test temperatures range from 15 to 35°C, but most seeds germinate well at about 20°C. The temperature may be steady or variable,

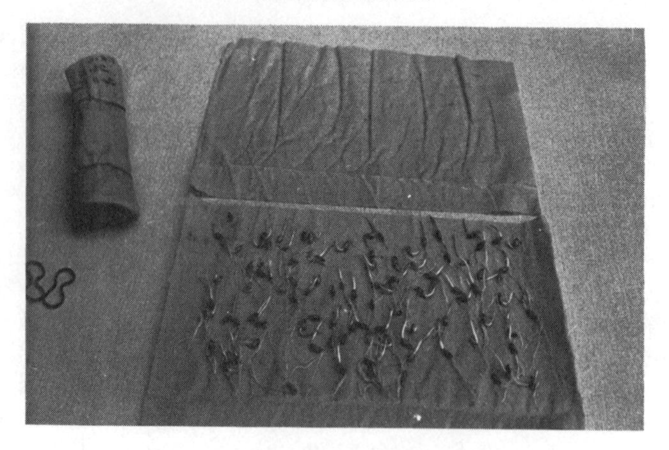

Figure 16.1 Barley seeds germinated in rolled paper towels (Scottish Agricultural Industries).

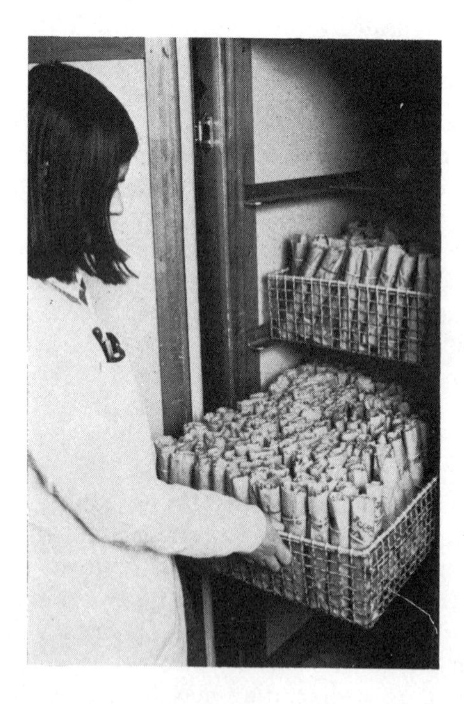

Figure 16.2 Rolled-towel tests ready for the germinator (Department of Agriculture and Fisheries for Scotland).

alternating between relatively high by day and relatively low by night. A common regime is 20°C for 16 hours and 30°C for 8 hours. This simulates the natural temperature fluctuations of night and day.

Alternating temperatures are necessary for dormant seeds only, but in some species, particularly grasses, dormancy is so common that this is the normal testing regime. The temperature change should be as rapid as possible.

Duration

The duration of the test is that which has been found by experience to give complete germination at the test temperature in the majority of samples. For the smooth working of a busy laboratory, the conventional periods should be adhered to, but occasionally the test can be extended by a few days if necessary. The period may be as short as five days or as long as 28 days, depending on the species.

The species which require a long test period are those in which dormancy is common. The sample is a mixture of dormant and non-dormant seeds, and germination is not uniform; non-dormant seeds germinate in a few days, and the dormant ones require weeks. For such samples, a *first count* is necessary after some days. The analyst counts and removes the seedlings which have developed sufficiently and would, if left to the end of the test period, grow excessively and possibly initiate decay.

Light

Among the grasses, dormancy is common and light is necessary to break it. In a regime of alternating temperatures, the warm phase should be in bright light, and the cold phase in darkness or dim light, as the combination of high temperature and darkness might induce further dormancy.

It is recommended, however, that tests at a steady temperature be carried out under strong light. In darkness or dim light the seedlings become long, slender and pale in colour, but in strong light they are short, sturdy and dark green, with the cotyledons spread out horizontally. Such seedlings look more natural, and it is easier for the analyst to judge whether they are normal or not. Another advantage is that if the seedlings remain short, a first count may not be necessary.

Breaking dormancy

When a sample fails to respond to the standard test conditions because of dormancy, special measures have to be taken. If at the end of the test period there are seeds which are not dead, but still firm and ungerminated, the test is repeated, either after pre-treatment, or with special conditions to overcome the dormancy. If at any time many samples of a particular species are found to show dormancy, as in the weeks following harvest, the pre-treatment or special conditions may be applied as a routine to all samples of that species. A common pre-treatment for cereals is chilling by keeping the sample for a few days on moist absorbent paper at a temperature of about 5°C in a refrigerator. Alternatively, barley and rice may be dried in a warm oven, or oats may be tested at a lower temperature than normal. Pre-treatment with gibberellic acid can break dormancy in cereals, and grass seeds can be induced to germinate by soaking the blotters in a dilute solution of potassium nitrate. In beet, dormancy is caused by a soluble inhibitor in the fruit wall, and this can be removed by washing in running water before the test. Dormancy in clover seeds is broken if the air is enriched with carbon dioxide and, if the moistened sample is enclosed in an impermeable polyethylene envelope, the carbon dioxide produced by the respiration of the non-dormant seeds is sufficient.

Equipment

Temperature is controlled by a thermostat set to maintain the seeds at the required temperature. Depending on climate and the time of year, both heating and cooling elements may be necessary. For uniformity of temperature throughout the germinator, it is necessary in some types to circulate air or water conditioned to the right temperature. Any light source generates heat, but fluorescent tubes less than others, and this has to be allowed for in the temperature control system. High humidity is necessary.

An arrangement which is effective in many situations is a small air-conditioned room with heat-insulated walls and a tight-fitting door. This room is maintained at a steady temperature by a heating or cooling unit, or both, and at a high humidity by a humidifier; to ensure uniformity, the air is circulated by a fan. Lamps are placed so as to give good lighting throughout. The temperature is set to suit the majority of samples being tested, and is kept constant. A steady temperature of 20°C serves temperate cereals, pulses and lucerne, for example, but many tropical crop species require a higher temperature. Test samples with the standard temperature requirement are placed on shelves. The minority of samples which have different requirements are placed in small germinator units within the room set to maintain a different temperature. The advantage of this system is that the difference in temperature that has to be maintained between the small germinator and the ambient atmosphere is reduced, and humidification within it is not necessary. There is therefore less load on the small heating units.

A cabinet germinator is a cubical unit with internal lighting and shelves to take the samples. The walls are double and contain heating or cooling elements. A reservoir of water maintains humidity, and air is circulated by a fan.

The Jacobsen germinator consists of a shallow tank holding water kept at a controlled temperature. It is covered by glass strips, by a perforated metal plate, or by a grid through which water circulates. The blotters are placed on it and kept moist by wicks dipping into the water in the tank (figure 16.3). Each blotter is covered by a transparent bell-jar, which reduces the water loss, but allows

Figure 16.3 Seeds germinating on a Jacobsen germinator (Department of Agriculture and Fisheries for Scotland).

ventilation through a hole at the top. The water is kept at the temperature necessary to maintain the seeds at the required temperature. If a rapid change of temperature is required, an automatic pump actuated by a time switch can empty the tank and refill it with cold water. Daylight can be supplemented by fluorescent tubes fitted above the tank. This apparatus is excellent for a temperate climate, but the blotters are liable to dry out, and it is not suitable for hot dry regions unless placed in an air-conditioned room.

Procedure

Table 16.1 is extracted from the International Rules for Seed Testing and indicates the methods to be used in testing seeds of some of the most important crop species. For any species, a choice of methods may be available. For example, maize may be tested on sand or in

Table 16.1 Germination methods for some important crop species, taken from the International Rules for Seed Testing.

Species	*Substrata*	*Temperature* °C	*Light*	*Duration* (days)	*Methods for breaking dormancy*
Avena sativa	S; BP	20	—	10	prechill; GA; test at 10–15°C
Hordeum vulgare	S; BP	20	—	7	prechill; predry; GA; test at 15°C
Oryza sativa	BP; TP; S	20–30; 30; 25	—	14	presoak
Pennisetum typhoides	BP	20–30	—	28	light
Sorghum vulgare	BP	20–30; 20–35	—	10	prechill
Triticum aestivum	S; BP	20	—	8	prechill; predry; GA; test at 15°C
Zea mays	BP; S	20–30; 25	—	7	
Phaseolus vulgaris	BP; S	20–30; 25; 20	—	9	light
Pisum sativum	S; BP	20	—	8	light
Vicia faba	BP; S	20	—	14	light; prechill
Lolium perenne	TP	15–25; 20–30; 20–25; 20	—	14	light; prechill; KNO3; test at 10–30°C
Poa pratensis	TP	15–25; 15–30 10–30	L	28	prechill; KNO3
Medicago sativa	TP; BP	20	—	10	prechill
Trifolium pratense	BP; TP	20	—	10	prechill; test at 15°C
Brassica oleracea	TP; BP	15–25; 20–30; 20	—	10	light; prechill; KNO3
Lactuca sativa	TP; BP	20	—	7	light; prechill; predry
Helianthus annuus	BP; S	20–30; 25; 20	—	7	prechill; predry
Corchorus capsularis	TP; S	30	L	5	

When more than one method is indicated, more than one may be used, but the first is to be preferred.
TP = top of paper; S = sand; GA = gibberellic acid; BP = between paper (including rolled towels); L = light essential.

rolled towels, at alternating temperatures of 20 and 30°C or at a steady temperature of 25°C. Which medium or temperature is selected depends partly on the materials and equipment available, and partly on past experience. If seeds remain fresh but ungerminated at the end of the test period, a second test has to be carried out in accordance with one of the dormancy-breaking techniques listed in the last column.

The test sample consists of 400 seeds taken at random from the pure seed fraction of the preceding purity test. The seeds are not placed all together, but in separate sub-samples, or replicates, usually of 100 seeds. The seeds are spaced out evenly on the sub-stratum, partly to ensure an equal supply of water to each, and partly to check the spread of any fungi that may develop. A great help in counting and placing seeds is a counting board for large seeds or a vacuum counter for small ones. The counting board consists of a tray with two layers, the upper layer with 100 holes and the lower layer removable to allow the seeds to drop on to the sub-stratum. The vacuum counter (figure 16.4) holds by suction one seed to each of 100 holes in a metal plate. When the plate is placed over the sub-stratum and the suction cut off, the seeds drop into place.

The replicates are placed in the germinator and (if possible) left until the end of the test period. In cases where uneven germination

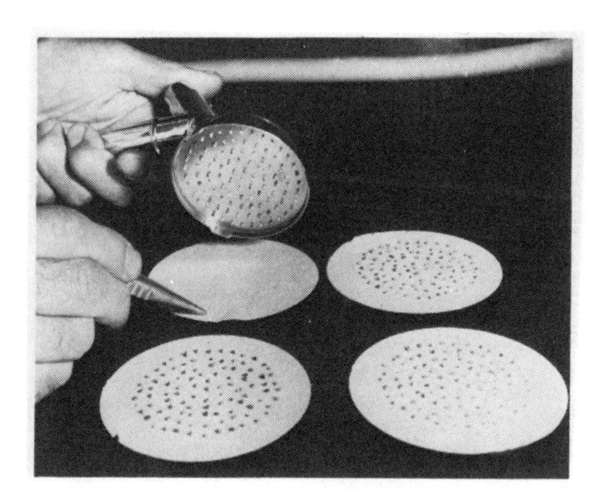

Figure 16.4 Vacuum counter for small seeds (Scottish Agricultural Industries).

necessitates a first count, any seedlings that have developed sufficiently to be evaluated are removed and recorded. At the same time, any seeds that are obviously dead can be removed and counted. The remainder are returned to the germinator.

At the end of the test period, each seedling is evaluated and classified as either normal or abnormal, and the numbers of dormant and dead seeds are counted. Dormant seeds are of two kinds—fresh ungerminated seeds and hard seeds. Fresh seeds have absorbed water, but show no signs of decay and may be expected to germinate if sown in the field. Hard seeds are common in small seeded legumes; they do not imbibe water and their sowing value in the field is doubtful. If the numbers of normal seedlings in the four replicates agree within statistical limits, and if there are not more than, say, five percent of fresh ungerminated seeds, the test is considered as finished and the percentage of normal seedlings is taken as the germination capacity.

Evaluation

Germination in a laboratory test is defined as the emergence and development of those essential structures which indicate the ability to grow into a normal plant in favourable field conditions. A seedling which displays these essential structures is said to be *normal*; a seedling with any defect is said to be *abnormal*. The essential structures are:

1. A well-developed root system.
2. A well-developed and intact hypocotyl or epicotyl (the stem growth above the cotyledons in hypogeal seedlings) and a normal plumule or, in cereals and grasses, a well-developed first leaf.
3. One cotyledon in seedlings of monocotyledonous plants or two cotyledons in seedlings of dicotyledonous plants.

Abnormal seedlings are of three general types—*damaged, deformed* and *decayed.*

Examples of damage are rootless seedlings in cereals and pulses (figure 16.5) and fractured hypocotyls in clovers. Most injuries are caused by rough treatment in threshing and subsequent cleaning, but can also be caused by rodents and insects in store.

Deformed seedlings have parts that are missing, misshapen or out of proportion. Examples in cereals are short thickened leaf sheath and roots, torn leaves and inverted embryos. Such abnormalities may be due to defects in the development of the embryo on the mother plant. This could be a genetic aberration, but more often it is due to some environmental factor such as bad weather after pollination or a mineral deficiency. For example, when there is a deficiency of manganese in the soil, seeds of peas are liable to show an abnormality known as "marsh spot", characterized by brown spots on the cotyledons and sometimes a plumule that fails to develop.

More commonly, the causes of these deformities affect the embryo after harvest. High temperatures during drying can produce defects in the seedlings, such as poor root growth and inhibition of shoot growth, particularly in cereals. In cereals again, an excessive dosage of pesticide produces a characteristic seedling with short thickened leaf sheath and roots (figure 16.6). Defects may also follow from storage in unsuitable conditions, such as high temperature or high humidity. When seed is subjected to prolonged storage, even in good conditions, the germination capacity eventually deteriorates. Some tissues remain physiologically alive, although the embryo as a whole has lost the ability to grow into a normal plant. The first symptom of old age is the appearance of abnormal seedlings in the germination test.

Decayed seedlings arise from seeds that have been infected by fungi or bacteria, which may be either pathogenic or saprophytic. An intact embryo is immune to saprophytic organisms, but these may gain entry if injury causes the death of any part of the embryo. A seedling is regarded as decayed only if it is clear that the decaying organisms

Figure 16.5 Normal (on the right) and abnormal seedlings of peas (Department of Agriculture and Fisheries for Scotland).

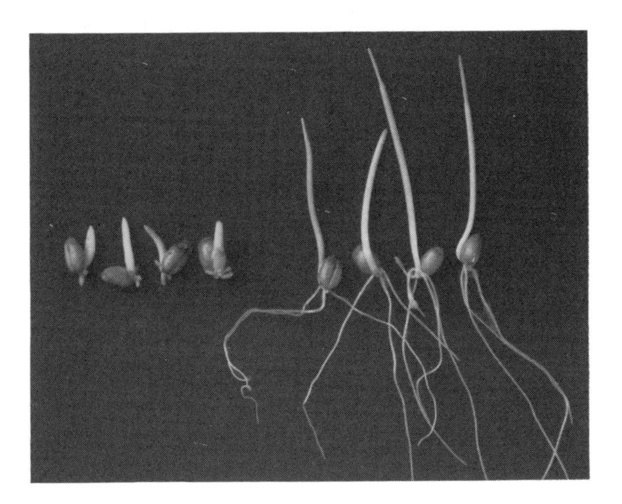

Figure 16.6 Abnormal seedlings of wheat showing typical symptoms of phytotoxicity due to an overdose of pesticide, with normal seedlings (on the right) for comparison (Department of Agriculture and Fisheries for Scotland).

have come from the seed itself and have not spread from another seedling during the test.

Sanitation

In carrying out germination tests, scrupulous cleanliness is essential. Many seeds are infected with moulds and other fungi, and these are liable to spread. Moreover, micro-organisms brought into the laboratory in this way are liable to set up a permanent source of contamination, capable of infecting future tests. After a test, therefore, all the germinated and dead seeds are destroyed, blotters and towels thrown out, and glassware, etc., thoroughly washed with detergent. Germinators are cleaned periodically by washing with a mild disinfectant, such as formalin or permanganate.

Sand is used repeatedly for tests and is sterilized by dry heat before use. Paper is used once only, but needs to be stored dry in packets which exclude dust.

Biochemical testing

An indication of the germination capacity can also be obtained by a biochemical test, using the reagent tetrazolium. Chemically, this is 2, 3, 5-triphenyltetrazolium chloride or bromide—either salt may be used. It is colourless and soluble, and diffuses readily into and through living tissue. There, in the presence of the enzyme dehydrogenase, it is reduced to triphenyl-formazan, which is red and non-diffusible, remaining inside the cells where it is formed. Chemically this is a test for the presence of dehydrogenase, but this enzyme is active in living tissue only so, in effect, it stains living tissue red and leaves dead tissue unstained. This reaction may also be brought about by light or, if the solution is alkaline, by other substances such as ascorbic acid. The solution of tetrazolium is therefore adjusted to a pH between 6.5 and 7 and kept in darkness both before and during the test.

In the test, the seeds are immersed in a 1 percent solution long enough for the tetrazolium to penetrate, and the staining of the embryo is noted. The intensity of staining is not significant. Embryos that are completely stained or completely unstained present no problem, but an embryo may be found to be stained in part only, indicating, for example, that the plumule is viable but the root is dead. In a germination test, the embryo is evaluated on the emergence of essential structures in the seedling; similarly, in a tetrazolium test, the embryo is evaluated on the staining of essential parts. The minimum staining required has been determined experimentally and is not the same for all species. For example, all the root tissue needs to be stained in sorghum, but only a small part in barley, and an intermediate proportion in wheat. For each species there is a minimum staining pattern which has to be learned by the analyst.

To facilitate the entry of the tetrazolium into the embryo and to make the staining pattern visible, preparation of the seed is necessary before immersion in the solution. This may consist of presoaking in water, removing the outer layers, or cutting each seed into halves.

If the germination capacity is high, the agreement between conventional and tetrazolium tests is good, but at lower germinations the tetrazolium test is less reliable. There are several reasons for this. Dormant seeds are not differentiated from non-dormant seeds. Abnormal growths may not be detected; embryos which develop into deformed seedlings may nevertheless stain normally. Living micro-organisms present in decayed embryos may cause staining. There is a risk, too, that in seeds killed by overheating in the drying process, or in seeds that have sprouted, dehydrogenase may persist for a short time and reduce the tetrazolium.

The great advantage of the test is its speed, and it is of particular value in processing establishments where incoming seed has to be assessed quickly, even at the risk of an occasional error. Normally the test is completed within 24 hours, but this time can be reduced by having the seeds take up the reagent under a vacuum, by maintaining a higher temperature during immersion, or by using the iodide salt of tetrazolium.

The test can also be used to decide whether seeds still ungerminated at the end of a conventional germination test are dead or dormant. The use of tetrazolium is not restricted to actual testing. It is a versatile tool and can be used to investigate the causes of poor germination or low vigour.

Vigour tests

The difference between germination capacity and vigour was discussed in chapter 1. *Vigour* has no precise definition, but in general terms it is the ability of a seed to germinate and grow into a plant in adverse field conditions. The critical point of difficulty is that adverse conditions are not always the same—in one country this may mean cold and wet, in another it may mean hot and dry. In both cases, the seed is placed under stress, but the stresses are quite different.

Another difficulty is that vigour cannot be given a precise numerical value. Among seed lots of the same species and with approximately the same germination capacity, all that a vigour test can do is to differentiate between those that will perform relatively well in poor conditions and those that will perform relatively badly. A vigour test is a supplement to the standard germination test.

Vigour tests have not yet been designed for all crop species, but those that are available are of two kinds—*direct tests* and *indirect tests*.

In a direct test, the adverse conditions which are expected when the seed is sown by the farmer are simulated in the laboratory; in this artificial environment the seeds are planted and the number of seedlings that develop are counted. It is difficult to standardize these conditions, so the figure obtained varies between one laboratory and another, and from time to time in the same laboratory. It is necessary, therefore, to plant a sample of known high vigour as a control to provide a comparison.

In an indirect test, measurements can be made of a physiological activity which is known to be associated and to vary with vigour, such as respiration, enzyme activity, and the permeability of cell membranes. In another kind of indirect test a possible cause of low vigour is looked for. Low vigour may be due to bruising of the tissues during processing, or to senescence, and these can be revealed by the staining pattern of a tetrazolium test. Other causes may be disease, which can be investigated in a health test, or small seed size, which becomes apparent in a determination of 1000-seed weight. Speed of germination (as observed in a germination test) is not a reliable indication of vigour because it is influenced so much by dormancy.

The commonest test methods are (1) the cold test, used in North America for maize, (2) the Hiltner test, used in Europe for wheat and rye, and (3) the conductivity test used in Northern Europe for peas. The first two are direct tests and the third is indirect.

In the cold test, the seeds are planted in moist soil containing *Pythium* and other fungi and kept at 10°C for a week, followed by three days at 30°C to complete the germination of seeds that are still viable. This subjects the seeds to two adverse factors—parasites and cold. In the Hiltner test, called after its inventor, the seeds are planted deeply in tightly packed brick dust. The adverse factor here is the physical force required for the seedling to reach the surface. When the permeability of the cell membranes in the seed tissues is faulty, soluble substances such as sugars and amino acids exude when the seeds are immersed in water and, in peas at least, this has been found to be associated with low vigour. In the conductivity test, measurements are made of the electrical conductivity of the water and this gives an indication of the amount of exudation.

Storage potential

In general, seed with a high germination capacity will retain its initial capacity in store for a longer period than seed of poor germinability. It does not follow, however, that seed lots of equal germination capacity will retain their capacity equally when kept in storage. The storage potential is influenced by such factors in the pre-storage history of the seed as weather during ripening and at harvest time, treatment during processing, and the conditions in which it has previously been stored.

A basis for forecasting the probable storage life of a seed lot is provided by the accelerated ageing process. This process aims at concentrating the stresses of prolonged open storage into a few days, by keeping a sample in the worst possible conditions without actually killing all the embryos. Suitable conditions are a relative humidity of about 90 percent and a temperature of 40–45°C, maintained for from two to eight days. At the end of this period of stress, the sample is subjected to a normal laboratory germination test. A high figure in this test indicates the ability of the seed to survive prolonged normal storage conditions. This is well illustrated by the following results from an experimental test of two contrasting samples of maize:

	Initial germination %	*Germination % after accelerated ageing*	*Germination % after 18 months open storage*
Lot A	97	92	92
Lot B	96	47	20

FURTHER READING

Delouche, J. C. & Buskin, C. C. (1973), "Accelerated ageing techniques for predicting the relative storability of seed lots," *Seed Science and Technology*, Vol. 1, p. 427.

McDonald, M. B. (1975), "A review and evaluation of seed vigour tests," *Proceedings of Association of Official Seed Analysts*, Vol. 65, p. 109.

Perry, D. A. (1972), "Seed vigour and field establishment," *Horticultural Abstracts*, Vol. 42, No. 2.

Wellington, P. S. (1970), "Handbook of seedling evaluation," *Proceedings of the International Seed Testing Association*, Vol. 35, p. 449.

ISTA (1976), "International Rules for Seed Testing," chapter 5, *Seed Science and Technology*, Vol. 4, No. 1.

Delouche, J. C. and others (1962), "The tetrazolium test for seed viability," Mississipi State University Agricultural Experiment Station Technical Bulletin 51.

17

TESTING FOR MOISTURE CONTENT, HEALTH AND UNIFORMITY

Determination of moisture content

The amount of water in a seed varies from time to time according to ambient conditions. The relationship of water to other components within the seed is complex, but it can be imagined as existing in three states, overlapping and not sharply distinguished from each other. Some of the water is an essential part of the living protoplasm and of the complex organic materials that constitute the endosperm. In this state it is chemically bound and is virtually independent of the seed's surroundings. In the second state, the water is held by electrostatic forces associated with the molecular structure of the seed. This water is held firmly, but less so than in the first state, and its amount varies to some extent. Thirdly, there is water in a state similar to the moisture in a wet blotter and which can be conveniently described as "free" water. Within the seed there may be some slight switching between the three states, but it is the free water which enters and leaves the seed according to the humidity of the atmosphere and affects its longevity in store. Theoretically, moisture content refers to the amount of this free water in the seed and it is expressed as the percentage of the total weight of the seed at the time the determination is made. This contrasts with the usual practice in the plant sciences whereby it is expressed as a percentage of the dry weight only.

The principle of a laboratory determination of moisture content is that the free water is driven out of a weighed sample, and the consequent loss in weight is taken to be the weight of such water that the sample contained. For agricultural seeds, the moisture is driven off by heat in a drying oven maintained at a constant temperature. The higher the temperature, the less time is required for drying, and for convenience this time should be as short as possible. It does not matter if the embryos are killed in the process, and the temperature can be above the boiling point of water.

At such temperatures most of the free water is driven off fairly quickly. The loss in weight of the sample is rapid at first and then becomes slow as the last of this water is removed. If the heating is continued beyond this point, more water is slowly lost, but this is bound water. There may be other substances present which are volatile and these, too, may be driven off at such temperatures. There is also the possibility of chemical changes, particularly oxidation, which would actually add to the weight of the sample. Oil seeds are specially liable to be affected in this way, but it depends on the chemical constitution of the oil. Linseed, for example, contains an oil which "dries" on exposure to air and this change is an oxidation which could occur in the drying oven.

For each species the appropriate drying temperature and time have to be found experimentally and strictly adhered to in routine tests. Ideally the temperature and time should be such that:

only free water is driven off.
nothing other than water is driven off.
no chemical changes take place.

Ideal methods take too long, and the methods recommended for routine laboratory use are not quite perfect. They may fail to meet one or more of these requirements completely, but they do give consistent and nearly accurate results. For this reason, moisture content could perhaps be defined more precisely as the loss in weight when a sample is dried under standarized conditions.

For agricultural seeds, two oven-drying methods are prescribed in the International Rules for Seed Testing—the low-constant-temperature method and the high-constant-temperature method. The low-temperature method involves drying at 103°C for 17 hours and is suitable for oily seeds such as soya, cotton and sesame. Most seeds, however, are non-oily, e.g. the cereals, pulses, grasses and clovers, and can be dried by the high-temperature method, which requires a temperature of 130°C and a time of 4 hours for maize, 2 hours for other cereals, and one hour for other species.

To ensure that samples dry completely in these times, large seeds are ground to a powder, unless they contain so much oil that they do not pass readily through a grinding machine. Seeds which should be ground include cereals and pulses. The machine must be capable of grinding quickly and without heating, so that moisture is not lost during the process.

If a sample is very wet, it will not grind down to a powder, and liquid water may even be squeezed out. Such seed is dried in two stages. In the first stage a large sample is weighed, spread out in a thin layer and left overnight in any warm dry place. After this predrying, the sample is weighed and the percentage loss of water calculated. This sample is then ground and a sub-sample subjected to the normal constant-temperature method. The moisture content is calculated by combining the water losses obtained in the two dryings.

Throughout the whole process, precautions have to be taken to prevent the moisture content changing on exposure to air. In the warehouse or store, the sample taken from the lot is placed in an air-tight container—a metal container with a tight lid, a glass jar with a screw top, or a bag made of thick plastic material and sealed with adhesive tape. The sample is sent to the laboratory for testing not later than the following day. In the laboratory, the seeds are ground, a sub-sample weighed out and placed in an already heated oven as quickly as possible. After drying and before weighing, the sample is allowed to cool in a desiccator kept dry by a desiccant such as phosphorus pentoxide.

These laboratory methods have been designed to give results as quickly as possible, but in practice there may be long delays, depending on the distance to the laboratory and the readiness of the analysts to test on receipt of the sample. The manager of a seed drying or processing plant, however, may require the result almost immediately, and in such circumstances an electrical moisture meter can be used. The meter measures an electrical property of the sample—the conductance or capacitance—which varies with the moisture content. The meter is calibrated for the species being tested, so that the dial reading indicates the moisture content. The calibration is checked periodically against a laboratory test.

These meters are most reliable in the moisture-content range of 10–25 percent, but they are not completely dependable in all circumstances, and may be inaccurate to the extent of one percent; temperature, for example, may affect the reading. They are essential tools for plant operators, to whom speed is more important than absolute accuracy; but when a seed lot is to be placed in long storage or issued for sale, its moisture content should be confirmed by the more dependable oven method in a laboratory.

An extremely simple way of determining moisture content is by the use of paper impregnated with lithium compounds. If a piece of this paper is enclosed in a bottle with the sample of seed, its colour after about two hours indicates the humidity of the air in the bottle, and from this the moisture content of the seed can be deduced with a margin of error. The method is not suitable for conditions where the ambient air humidity is high.

A modification of the oven method is a simple piece of apparatus which utilizes an infra-red lamp as a source of heat and can be used in a store to give quick results. It can also be used when the moisture content is too high or too low for an electric meter but, because of its high temperature, it is not suitable for seeds with a high oil content. It is less sturdy than a meter, and is liable to deteriorate in warehouse conditions.

As for all seed attributes, the test result applies to the sample and is only applicable to the lot insofar as the sample is representative. Within a seed lot there may be considerable variation in moisture content and, if the test is to be of any value, the sampling procedure must be followed meticulously. Even when the result is a true indication of the average moisture content of the lot, it should be borne in mind that there may be spots within the lot where it is appreciably higher.

Testing for seed-borne diseases

Seed-borne diseases are caused by fungi, bacteria, viruses and eelworms. Different diseases require different laboratory methods for their detection. It is not possible to identify by one test all the pathogens that might be present (and it would be quite unreasonable to expect a sample to be tested for freedom from all disease) but, if certain pathogens are specified, it is possible for a pathologist to say which of them is present and to what extent.

Some diseases can be detected directly by examination of the dry seeds with low magnification or even with the unaided eye. Fungi which produce hard dark-coloured sclerotia or fruiting bodies can be detected in this way, e.g. ergot in cereal seeds, *Sclerotinia* on clover seeds, and *Septoria* on flax and celery seeds. Discoloration caused by bacterial diseases of pulses and galls produced by eelworms can also be seen.

Another quick technique is to shake up a seed sample in a little water and examine the suspension microscopically for any evidence of micro-organisms that have been washed off the surface. This method is particularly useful for certain smut and bunt diseases of cereals; these are caused by fungi which do not grow in or on the seeds in the sample, but the resting spores are carried mechanically on the surface.

In the case of loose smut, however, the fungal mycelium is actually present within the tissues of the embryo; and for the loose smut of barley, a method has been developed of separating the embryos from the other seed tissues and examining each embryo microscopically. If the pathogen is in the superficial layers of the seed, it is possible to detect it by less complicated methods.

In the blotter test, seeds are placed on moist blotters (as for a germination test) and maintained in high humidity at a temperature between 20 and 30°C. This provides favourable conditions in which both the seedling and the pathogen can grow. After a few days the pathogen can be recognized by the symptoms that appear on the seedlings or by the fungal growth that emerges. Oat seeds infected with stripe disease (*Drechslera* spp.) produce seedlings with brown stripes on the first leaf. Cereal seeds infected with root rot (*Fusarium* spp.) give seedlings with brown roots. A modification of this method is to prevent the seedling from developing by deep freezing for 24 hours or by treatment with the herbicide 2,4-D. The pathogen is not affected and becomes obvious on the surface of the seed. The species of fungus is identified by its spores, and spore formation is promoted by exposing the seeds to ultra-violet light.

Another method is to place the seeds on the surface of a sterile nutrient agar jelly in a Petri dish. Any fungus present grows out from the seed and forms a colony on the agar, and the fungus can be identified by its colour and form of growth. Before being placed on the agar, the seeds are washed with a dilute solution of hypochlorite. This kills any fungal spores accidently adhering to the surface, and ensures that any fungus which grows out into the agar was present internally. Even so, there may be saprophytes present, and the colonies which form have to be identified as pathogenic species.

For some diseases there is no test that can be completed in a few days. All that can be done is to plant the seeds in favourable growth conditions in a glasshouse and await the development of symptoms in the plants that emerge—and this may require several weeks. An example of a disease that can be identified only in this way is lettuce virus.

Bacterial diseases can be identified by special bacteriological methods, using phage and serological techniques.

Testing for uniformity

As explained in chapter 14, the maximum size of a seed lot for quality control and labelling purposes is either 10 or 20 tons, depending on the size of the seed; and this is sold to farmers in packages of 50 kilograms or less, so that a single lot may comprise more than 400 packages. It is desirable that within a lot the contents of each package should be the same, but in practice complete uniformity is rarely attained. The amount of variation between packages can be measured by a test for uniformity.

For this test, separate samples are taken from a number of packages—up to 30, depending on the size of the lot. Each sample is tested for one attribute—the percentage of pure seed, the germination capacity, or the number of weed seeds. Whichever attribute is tested, there will be some variation in the values obtained. A mathematical measure of the variation is easily calculated and is called the *variance V*.

This is compared with the variance that would be expected theoretically if the variation between sample values were entirely at random *(W)*, i.e. if the pure seed, or the germinable seeds or the weed seeds, were distributed at random throughout the lot. The standard for comparison is not complete uniformity, in which the variance would be nil, but a state of randomness in which uniformity is less than complete.

The comparison is made by calculating the ratio

$$\frac{V}{W} = \frac{\text{actual variance found}}{\text{variance to be expected in a state of randomness}}$$

If the pure seed, viable seed, or weed seeds were distributed at random, the ratio would be unity, but usually it is greater. From this ratio the *heterogeneity value* can be obtained by subtracting one,

$$H = \frac{V}{W} - 1$$

This means that the standard of uniformity is taken to be that in which variation is at random, and the H-value is an indication of the excess of the actual variation over this standard. As randomness is not uniformity, the actual variance is sometimes less than the standard. In this case the ratio is less than unity, and H has a negative value.

A low H-value does not necessarily indicate that the lot has been well blended. If a seed lot were made up by combining two or more batches which had been so thoroughly cleaned that no weed seeds remained, then a uniformity test made by counting weed seeds would give a low H-value even without any blending. A high H-value, however, is a sure indication that blending has been inadequate.

This test is not practicable for routine purposes because of the excessive amount of work involved in drawing and testing so many samples. It is, therefore, not possible to set minimum standards of uniformity which must be complied with. The test is of value, however, in checking the efficacy of blending methods, and for this purpose should be known and used by plant managers.

FURTHER READING

Thomson, J. R. (1972), "The heterogeneity test," *Proceedings of the International Seed Testing Association*, Vol. 37, p. 669.

FAO (1975), "Cereal Seed Technology," Appendix to chapter 7, Agricultural Development Paper No. 98.

ISTA (1976), "International Rules for Seed Testing," chapters 2, 7 and 9. *Seed Science and Technology*, Vol. 4, No. 1.

18
LEGISLATION

A LAW GIVES GOVERNMENT POWER EITHER TO DO CERTAIN THINGS OR TO oblige its citizens to follow (or refrain from) certain procedures. The primary purpose of such legislation as applied to seed is either to protect the interests of individual farmers or to improve the agricultural productivity of the country—but, of course, both purposes can be served by the same law. The effect is to promote the use of high-quality seed by farmers.

Following the pioneer work of Professor Nobbe last century, investigations were carried out into the quality of seed being used in European countries. These investigations showed that much of the seed being sold to farmers was of poor quality, either fraudulently or due to ignorance on the part of both seller and buyer. Early legislation was inspired by revelations of this kind, and was explicitly intended to protect the farmer. This justification for seed laws still exists to some extent, depending on the stage of development of a country's agriculture and the sophistication of its farmers, but legislation also protects reputable seed merchants against unscrupulous competition.

In Britain various agencies undertook the testing of seeds, but legislation had to await a national need. During the war of 1914–18 the import of food was restricted by a submarine blockade, and the maximum use had to be made of the country's own farming capacity. It was then realized how much production was being lost through the sowing of seed which was either of the wrong kind or was incapable of germinating. Emergency regulations were therefore introduced, requiring all seed to be tested before sale. This was so effective that after the war the legislation was given permanent form in the Seeds Act of 1920, and this continued unamended until 1964, when it was superseded. Legislation in Asian countries has been strongly influenced by the need to improve agricultural productivity.

Much of the seed that is sown is either saved by the farmer himself from a previous crop or obtained from a neighbour. It is therefore quite impracticable to ensure that all the seed which is sown is of high quality, and seed laws concentrate on controlling the seed that is publicly sold to farmers.

If the purpose of seed laws is to protect the farmer against the unwitting purchase of poor-quality seed, why is there not similar legislation covering goods such as pots and pans, nuts and bolts, trousers and shoes? In some countries there are indeed laws protecting the buyer of such goods, but it can be argued that there is a fundamental difference between living seeds and inert household utensils. The buyer may be able to judge the quality of hardware and its suitability for his purpose before purchasing. This, however, does not apply to seeds; he cannot in many cases confirm the species or recognize impurities; he certainly cannot identify the cultivar or assess the germination capacity. There is, therefore, ample opportunity for fraud and deception but, even if the seller is an honest man, without a laboratory test or a certification scheme, he is no better placed than the buyer. Another significant difference between seeds and other goods is that in the case of hardware the greatest loss likely to be incurred by the buyer is the cost of the goods, but a farmer sowing seed of poor quality may lose the whole crop and possibly his livelihood for a year.

Seed laws can achieve their aim only if there is high-quality seed available to be promoted. Legislation alone produces no seed. Furthermore, there must be recognized channels of trade along which seed passes to farmers, and an administrative system which is capable of controlling these channels. In other words, it is premature to initiate seed legislation until there are known sources of good seed and enforcement can be effective.

Every national government has an agricultural policy which it pursues in compliance with the laws passed by the legislature. Only part of this policy is concerned with seed, and every enactment which has implications for seed is not necessarily included in a specific seed law. The following are some examples of government activities which can have a bearing on the quality and availability of seeds, but which might not be authorized in special seeds legislation:

Agricultural research and plant breeding
Extension and advisory services to farmers
Agricultural education
Quarantine
Control of export and import trade
Price control
Food rationing
Health and safety precautions

Legislative strategies

There are two types of seed legislation based on "truth in labelling" and "minimum standards".

Where legislation is of the former type, there is no ban on the sale of seeds, no matter how worthless they may be. A seller is, however, required to give the buyer certain information by means of a label or otherwise. This information typically includes the name of the species and cultivar, analytical purity, weed content and germination capacity. It may be based on an official laboratory test or certification scheme, or it may be entirely the seller's responsibility. The essential point is that it must be true at the time of sale. Enforcement is carried out by inspectors visiting warehouses and shops, and taking samples from the seed lots ready for sale. These samples are tested at an official seed-testing station and, if the tests show the information on the label to be inaccurate, sale of that lot is stopped and the owner becomes liable to prosecution.

Under "minimum standard" legislation, standards of seed quality are decreed for such attributes as analytical purity, weed content, germination capacity and possibly health. All seed has to be tested by an officially recognized laboratory, and is approved and labelled for sale only if it complies with these standards. The seeds may have to be of an approved cultivar and certified under an official scheme. Enforcement is by means of inspectors visiting warehouses and shops to check that seed exposed for sale bears official labels and to take samples for confirmatory testing.

Which type of legislation is more appropriate for any particular situation is a matter for judgment in the light of local circumstances. Truth in labelling is suitable for sophisticated farmers able to understand the information given and assess its importance, and who are economically capable of exercising some choice. Minimum standards are more suited to small peasant farmers who may not be able to appreciate the information on the label or, being dependent on one source of supply, are not in a position to reject what is locally available.

As a seed law can serve no useful purpose until there is good seed available, it follows that a vital function of the law is to provide the means by which good seed can be distinguished from bad. One or more seed-testing stations have to be set up to provide a testing service, not only for enforcement officers, but also, for their own information, for farmers, merchants, processors and everyone concerned in the production of high-quality seed.

Acts and Regulations

A complete seed industry is not organized instantaneously, but takes years to develop. It is advisable, therefore, for the law to develop on parallel lines, not seeking to control what does not exist or to improve procedures or standards which cannot be enforced. So the law should apply only to those crop species of which high-quality seed is available; as good seed of other species becomes available, these can be brought within the ambit of the law. At first the law may be applied only in the best farming districts of the country, leaving poorer or more remote areas until later. Initially the law may concern only those elements of seed quality which can be tested in a laboratory; schemes to certify cultivar purity can follow, on a voluntary basis at first, perhaps becoming compulsory later.

The procedure for enacting a law or amending it is cumbrous and time-consuming, but it is not necessary for this procedure to be repeated as the industry develops or as technological advances are made. Given a broad vision of how it is expected to develop, this difficulty can be overcome by the device of having two kinds of legislative instruments—an Act with Schedules, and Regulations—and in the Act provision is made for simple and rapid ways of amending the Schedules and Regulations.

The Act is enabling; it lays down the general principles and declares the objectives. It describes, without going into details, how these objectives are to be achieved—seed is to be tested for certain attributes and may be certified as to cultivar, cultivars are to be approved and listed, growers and sellers may have to be registered and to keep records, seed packages are to be labelled, enforcement is to be through sampling, but tolerances are to be allowed, and so on. The terms used are defined, such as *breeder, basic seed* and even *seed.* Inspectors are authorized to enter stores and shops, to draw samples and to stop further sales of inferior seed; procedures for prosecution and penalties are prescribed. Matters which may have to be amended from time to time are included in Schedules, e.g. the kinds of seed to which the Act applies, and the quality attributes to be controlled. None of this, however, can become effective until Regulations are drafted and promulgated, and the Act gives the Minister of Agriculture authority to do so, within specified limits. These

Regulations have the full force of law and they can be amended without going through the full process of legislation, though the legislature can, if it wishes, intervene. The Schedules in the Act can be similarly amended.

The Regulations supply all the details—what weeds are to be regarded as noxious, how samples are to be drawn, the precise standards to be observed (if any), the tolerances that are acceptable, arrangements for the assessment of cultivars and for certification and testing, information to be stated on labels, sealing requirements, how imported seed is to be controlled, and so on. Certification authorities and cultivar approval boards are set up. Some exemptions are usually considered necessary, such as seed for experimental purposes and seed which is purchased with a view to processing before sale to farmers.

How the law works

For quality attributes which can be assessed in a laboratory test, the procedures for complying with the law and for enforcing it are simple, and samples for test are taken from the seed which is to be sold. In a "minimum standard" regime, labels are attached at the time of sampling, and the bags are sealed to prevent tampering.

To control cultivar purity, an official certification scheme has to be operated. This is managed by a Certification Authority which may be the Ministry of Agriculture or a board with nominated membership drawn from interested groups such as farmers, seed merchants, co-operatives, scientists and extension workers. A certification scheme is operated on the lines indicated in chapter 12.

Approval of cultivars is based on tests of distinctness, uniformity, stability and usually on trials of usefulness (see chapter 13). Trials of agronomic value have to be carried out on numerous locations scattered over the country, and finding suitable sites may be a difficult problem; but the number of centres required for determination of distinctness, uniformity and stability is relatively small. This complex of trial activities is controlled by a board charged with the responsibility of management and of advising the Minister as to which cultivars should be approved. The members nominated to this board may include agronomists, biometricians, extension workers, representatives of consumer interests, trade organizations, credit agencies and of the certification authority. Cultivars produced by foreign breeders are usually eligible for approval equally with home-bred cultivars.

Approved cultivars are registered in a statutory list for a fixed period in the first place of, say, 10 years. A cultivar may be removed prematurely if it fails to fulfil its early promise, or have its registration extended if it is still contributing substantially to agricultural productivity. Along with each cultivar, the name of the breeder who is responsible for its maintenance is registered. Maintenance involves conservation of the parental material and the production of basic seed; the maintainer is normally the original breeder of the cultivar or his authorized agent.

A statutory list becomes a long inventory of names without any indication of special merits or relative value. Some of the cultivars are suitable only for certain districts or for a particular purpose; some of them, though still popular with farmers, become outclassed by newer products. The statutory list, therefore, may well be supplemented by a shorter list of recommended cultivars issued by the extension service. Such a list can indicate relative merits, special features and regional suitability, but it is purely advisory and has no legal status.

It is not necessary for all seed to be certified as to cultivar, and there may be a non-certified category designated as "commercial", which is controlled by laboratory tests only. In some crop species there may be no provision for certification, and all seed is sold as commercial; this usually applies to vegetables and minor agricultural crops. In other species, certification may be voluntary; there is an official scheme, but it is not obligatory, and growers and merchants may submit to the scheme only if they wish. Uncertified seed is sold under the commercial category and is clearly labelled as such.

Law enforcement officers should not be regarded merely as policemen seeking out farmers and merchants who contravene the seed laws, in order that they may be punished. They should know enough about the law and its objectives to explain its requirements and be able to advise how the law can be complied with, e.g. by proper storage or having samples tested. In other words, they should have an extension and advisory function.

Cultivar names

Before a cultivar is listed, its name has to be approved. The main reason for this is to avoid confusion with other cultivars. It would be quite unacceptable to have two cultivars with the same name within the same crop species, but it is advisable to extend this restriction to crop species of the same type, e.g. Italian ryegrass and perennial ryegrass, or cowpea and *Dolichos*.

A name which is a registered trademark should be avoided, as this might enable the breeder to prevent other people using the name and thus secure a monopoly of the cultivar.

As a guide to authorities who have to approve cultivar names there is the International Code of Nomenclature for Cultivated Plants. This was prepared by a commission of the International Union of Biological Sciences which, under the aegis of the United Nations Educational, Scientific and Cultural Organization (UNESCO), promotes international collaboration in scientific matters.

The International Code of Botanical Nomenclature, issued by the International Botanical Congress, governs the use of botanical names in Latin form. The code for cultivated plants governs the use of non-Latin names for cultivars within a botanically named species. It aims to encourage uniform practice in naming cultivars by stating guiding principles and making specific recommendations about the formation, use and recognition of such names. The code has no legal status, but in practice it is voluntarily followed by many governments in approving and registering cultivar names. Precision and uniformity are particularly necessary in connection with plant breeders' rights.

The following are some of the recommendations included in the code:

Numerals and symbols should not be used as names, e.g. "S.23" perennial ryegrass.

Names should not exaggerate the merits of a cultivar, as they may become inaccurate through the introduction of new cultivars, e.g. "Earliest of All" tomato.

Names should consist of not more than two words.

Names should be simple, short, easily pronounced and unlikely to be misspelt.

After a cultivar has ceased to be grown by farmers, its name should not be used for another cultivar for at least ten years.

Plant breeders' rights

Closely related to the laws which directly control the sale of seed, are laws concerning the "ownership" of cultivars.

The first improved cultivars were introduced by individuals, either working for themselves or employed by firms of seedsmen. It was not until Mendel's laws of inheritance were rediscovered early this century (see chapter 1), that scientific institutions became interested. Applying their newly acquired knowledge, scientists working in universities and research institutes produced new cultivars which were made available to farmers. This new knowledge was no secret; it was published in scientific journals and taken up by commercial seedsmen who also produced new cultivars and marketed the seed. In some countries, such as Germany and France, scientific institutes tended to concentrate on fundamental research, leaving the production of cultivars for the agricultural industry to the seedsmen, and some of these developed large plant-breeding stations as part of their commercial activities. In other countries, of which Britain and Sweden were notable examples, the scientific institutes took their breeding activities further and produced cultivars for direct release to farmers.

In the beginning, producing a new cultivar was, in certain crop species, extremely simple. In self-fertilized species like wheat, the farmers' stocks were very mixed, and all that was necessary was to select a single outstanding plant, save the seed from it, and multiply it over a few generations to produce a pure-line cultivar. In time, however, the possibilities of obtaining significant improvement by simple selection were exhausted, and much more complex procedures were developed involving extensive screening, complex cross-pollinations, chromosome manipulation, and so on. Thus plant breeding became more costly; publicly financed institutions were able to obtain more money from the taxpayer, but commercial breeders found it increasingly difficult to recover their costs through the sale of seed.

The crux of this problem lies in the maintenance of a cultivar after its initial creation. This involves conserving the parental material, and from it initiating each year an issue to be multiplied over a number of generations to produce basic seed. In a cross-pollinated species this may be quite difficult, and only the breeder with his intimate knowledge of the genetic make-up of the cultivar and the variants that are liable to arise, is capable of doing it adequately. In the case of an F1 hybrid cultivar, if the breeder has himself produced the inbred parental lines, nobody but the breeder can produce seed of the cultivar, and he has a monopoly.

In the case of cultivars which are predominately self-pollinated, no special knowledge is necessary for maintenance. Care is certainly needed, but any competent seedsman or agronomist can maintain such a cultivar for several generations without serious degeneration. In a free market, therefore, when a new self-pollinated cultivar is released, seed can be bought and multiplied for sale by any seedsman—and the breeder loses control. He can only recoup his development costs by charging high prices for his original release and for the recurring demand for basic seed to enter a certification scheme, if there is one; sales of basic seed depend on the number of certified generations that are permitted.

So it came to be realized that if commercial breeders were to continue making their special contribution to agricultural progress, some means must be devised, not only to recoup their costs, but to provide an incentive for further breeding work. A breeder can not be guaranteed against loss, but his reward should be in proportion to the extent to which his cultivar is grown by farmers, which still leaves a degree of commercial risk. It was also thought necessary to encourage breeders to maintain their cultivars for the production of basic seed, rather than to abandon them when they became unprofitable.

Various systems of rewarding breeders were tried in different countries. Cultivar names were registered as trade marks, thus restricting their use to the breeders, but this protected the name rather than the cultivar (which could be sold under another name) and tended to limit the general acceptance of a cultivar. Other countries tried various ways of restricting the number of cultivars available for sale and paying subsidies to the breeders whose cultivars were listed, but this did not allow the free play of market forces in determining a breeders' reward.

Most countries have patent laws to protect inventions, the protection covering the method and the product, and it was thought that it might be possible to protect cultivars in the same way. It was objected, however, that a mechanical or material invention could be precisely described and defined in words, whereas a cultivar was a living population liable to show genetic variability. Moreover, a patent protected the method of manufacturing the product, but except for F1 hybrid cultivars, creation of a cultivar is done once only and is never repeated, and methods of maintenance are common knowledge.

The concept developed, therefore, of a cultivar as an entity in itself which could be protected by controlling its use. Protection would be more like the copyright of musical and literary works than patent rights. A musical composition is created once, but it is not freely available to the public. The composer retains controlling rights, and it can only be performed in public with his permission, which he may grant on conditions, such as the payment of a royalty fee. Similarly, a cultivar is created once only, and the breeder could have the right to control multiplication and sale of seed, and to demand royalties.

At meetings held in Paris in the 1950s it was agreed by representatives of several European countries that this concept of plant breeders' rights should be accepted, and the International Convention for the Protection of New Varieties of Plants was eventually negotiated and signed in 1961. Under this convention, the signatory countries undertook to introduce legislation on the lines laid down in order to enable breeders to secure legal rights in their cultivars. This convention has so far been ratified by less than a dozen countries, mainly European, but is open to all states. The legislative action taken by a signatory country can provide for rights within that country only, but breeders in other countries are eligible to apply for these rights.

To be eligible for protection, a cultivar must satisfy four conditions:

(i) It must not have been commercialized previously.
(ii) It must be distinct and clearly distinguishable by important characters which are capable of precise description and recognition.

(iii) It must be uniform to a degree depending on the breeding system of the species.
(iv) It must be stable and remain true to its description after repeated reproduction. In the case of F1 hybrid cultivars, this means that a cultivar must be true to its original description each time it is recreated by hybridization.

Evidence of merit in field performance and product value is not required, the assumption being that only cultivars good enough to become popular deserve rewards. Protection is granted after the cultivar has shown in official tests and examinations that it does in fact meet the requirements of distinctness, uniformity and stability.

The cultivar must be registered under a name which is unique within its genus and will neither mislead nor confuse, and no other name may be used. It must not be a trade mark. The protection lasts for a period of not less than 15 years.

The nature of the rights is defined in the convention. The essence is that the breeder's prior authorization is required for the production of seed for sale, but not by a farmer for his own use. In practice this means that the breeder can restrict the multiplication for seed of his cultivars to persons licensed by himself, and may require persons so licensed to observe certain conditions such as the payment of royalties. Royalty payments may be calculated on the quantity of seed sold or on the area of land used in multiplication; the crop area officially certified, for example, provides a convenient measure. The government is not actively involved in enforcing these rights; they are civil rights which the breeder must pursue himself through the civil courts, i.e. if a seedsman sells seed without paying royalty, he can be sued in a civil court, but he will not be prosecuted in a criminal court. Nevertheless, government may be anxious to prevent the creation of monopolies and intervene to ensure that applications for licenses are refused only with good reason.

The United States has not signed the convention, but has introduced a system of rights which differs in some respects. The most important difference is that the breeder is required to submit a detailed description of the cultivar on a prescribed form and a sample of seed which may be used to confirm his description. He is expected to carry out tests and examinations on which his description can be based. All descriptions are recorded in a computer and, in practice, the granting or refusal of protection is a computerized operation.

The International Union for the Protection of New Varieties of Plants (UPOV) is an organization with headquarters at Geneva, formed by the countries which adhere to the convention, in order to co-ordinate the administrative and technical procedures followed in member countries for the granting of rights. It is possible for a cultivar to be accepted in one country, but rejected in another because of differences in the testing techniques or in the standards expected.

The Union's principal effort so far has therefore been directed to the publication of guidelines for the conduct of tests. These consist mainly of lists of characters which should be examined in each species. The ultimate aim is to have each cultivar tested by one institution only, and for each country to accept its verdict. This, however, may turn out to be no more than an administrator's dream. With the vast increase in the number of cultivars, the differences between them have become very fine, and the expression of certain distinguishing characters is influenced by day-length; distinctness and uniformity can therefore be modified by latitude.

There is no doubt that in Europe the introduction of plant breeders' rights has given a tremendous impetus to plant breeding. Commercial breeders have confidently launched ambitious programmes, and the number of cultivars submitted has been an embarrassment to the testing authorities. For the successful breeders it is an exceedingly lucrative business. But the farmers have benefited too.

FURTHER READING

FAO (1975), *Cereal Seed Technology*, chapter 9, Rome.

International Association for Plant Taxonomy (1969), *International Code of Nomenclature for Cultivated Plants*, Utrecht.

ISTA (1967), "Seed Legislation," *Proceedings*, Vol. 32, No. 2.

UPOV (1974), *International Convention for the Protection of New Varieties of Plants*, Geneva.

19

PLANT NAMES

LANGUAGE IS MADE UP OF ALL THE WORDS THAT MAN HAS DEVELOPED for the objects, actions and emotions of his life. The words used by early man for plants were not very precise—grass, corn, beans and so on—each one covering a wide range of plants, though all with something in common. The need for precision in plant names arose originally in medicine; until last century, remedies for disease were herbal in origin, and a mistake in harvesting the plant material could have fatal consequences.

The science of botany started with the collection and description of different kinds of plants. In Europe, the early botanists, living in different countries, wrote their descriptions in Latin because that was the international language of the time, through which scholars in all philosophies communicated with each other. Attached to each description of a kind of plant there had to be a scientific name for it, which botanists could use. Before being studied by scientists, many wild plants had no special names, while others had different names in different countries, and the use of a name which was associated with a published description prevented confusion. The scientific names, like the descriptions, were in Latin; they may not all have been words actually used in Rome 2000 years ago, but they were in a Latin form, as they still are today.

Plants are named in accordance with the International Code of Botanical Nomenclature, which was established by an international convention of botanists. This code lays down rules on such matters as the form the name should take, which name takes precedence if in the past a plant should have been given more than one name, and the procedure for changing names. Names may have to be changed in consequence of additional knowledge acquired in scientific research.

Agricultural research is now international; scientists exchange seeds and eagerly scan the results of investigations carried out in other countries for information that can be used in increasing productivity. For this movement of material and knowledge to be effective, the plant names used must have precise meanings understandable by scientists all over the world. Furthermore, laws designed to prevent the international spread of noxious weeds, must give these weeds names which are understood in all countries. So it is generally accepted that plants should be named in accordance with the code, but nevertheless there is some doubt as to the validity of hundreds of plant names.

To deal with this situation, the International Seed Testing Association has published a list of stabilized plant names and recommends that in cases of doubt the listed names should be used. The inclusion of a name in this list is not a judgment on the validity of the name; it is a recommendation of convenience to be followed until scientific research has resolved the doubt. Additions to this list are published from time to time.

What is a species?

In classifying and naming plants, the basic unit is the species, but a species is difficult to define. For each species there is a description written by the botanist who first named it, and dried specimens of the plants are preserved in the herbarium of some botanic garden. The original descriptions were based on the specimens, though often supplemented by examination of living plants.

A species is distinctive and consists of all the plants which individually fit the description and are similar to the type specimens. This does not mean that all the plants in the species are identical, because the descriptions usually allow for some variation. For example, the description of subterranean clover includes the following items:

More or less clothed with long spreading hairs.
Stems *usually* short, but *occasionally* lengthened to 6 *or* 8 inches.
Flowers white *or* pale pink, two *or* three together.

How much variation is permissible between plants within a species is not prescribed in the code, but is a matter for judgment by botanists experienced in this special field of work. It depends to some extent on how many plants have been examined and from how wide an area. A species may be described and named on the evidence available at the time, but later investigations may show that it can be divided into two. Alternatively, evidence may accumulate that the species is not as distinct as was thought, that intermediate forms exist, and that it should be amalgamated with another species.

In cultivated species this variation has been accentuated by the selection of different types of plants to suit the special needs of farmers and growers. Distinct populations of plants known as *cultivars* have been built up, each cultivar being uniform for certain characteristic features of importance to the farmer or user.

In general, a plant will set seed if it is fertilized by pollen from another plant within the species, but will not cross with a plant of another species. This means that the characters of the species are transmitted from generation to generation within the species, and are not transferred to other species. So a species remains distinct, and hybrids between species rarely arise.

Specific names

A number of broadly similar species is grouped together to form a *genus* (plural *genera*). Genera are distinguished from each other by major differences, mainly in floral structure; within a genus, all the species share the characteristics of the genus, but differ in minor characters, often in the stems and leaves.

A species is referred to by two names: first the name of the genus, followed by the name of the species, and each is in Latin form. This system was first used by Linnaeus, a biologist who lived in Sweden about 200 years ago. The species name may be followed by an initial or an abbreviation to indicate the name of the botanist who described and named it. Very many names are followed by "L.", which stands for Linnaeus.

A good example of the use of generic and specific names is provided by the wheats. All the wheats are grouped together in the genus *Triticum*, which is the Latin word for wheat. Within this genus are several species, and each has been given a name which is in some degree descriptive, as follows:

Triticum aegilopoides	is the wheat like aegilops grass.
Triticum monococcum	is the wheat with one grain in each spikelet.
Triticum dicoccum	is the wheat with two grains in each spikelet.
Triticum durum	is the wheat with hard grains.
Triticum sphaerococcum	is the wheat with spherical grains.
Triticum turgidum	is the wheat with swollen (turgid) glumes.
Triticum compactum	is the wheat with compact ears.
Triticum polonicum	is Polish wheat.
Triticum vulgare	is ordinary wheat.

As many specific names are descriptive, it is of interest to know what they mean, and a few of them are listed below as examples. The last syllable is liable to be altered in accordance with Latin grammar.

arenarius	growing in sandy places	*caespitosus*	tufted
palustris	growing in swamps	*erectus*	erect
pratensis	growing in meadows	*repens*	creeping
campestris	growing in fields	*hypogaeus*	underground
sylvaticus	growing in forests	*grandis*	large
sativus	cultivated	*major*	larger
esculentus	edible	*minor*	smaller
medicinalis	used in medicine	*medius*	intermediate
sylvestris	wild	*minus*	small
trivialis	common	*annuus*	annual
lunatus	crescent-shaped like the moon	*perennis*	perennial
ovatus	oval-shaped	*aestivus*	summer-flowering
dentatus	toothed	*vernus*	spring-flowering
glabra	smooth	*praecox*	early
inermis	without thorns	*rubus*	red
hirtus	with short hairs	*niger*	black
hirsutus	with long hairs	*senescens*	grey, like an old man
molle	with soft hairs	*viridis*	green

The scientific Latin names of the plant species mentioned in this book are listed in Table 19.1 along with the names commonly used in five different languages.

Table 19.1 Latin names of plant species and the names commonly used in different languages

Latin	*English*	*French*	*German*	*Spanish*	*Arabic*
CEREALS					
Avena sativa	oats	avoine	Hafer	avena	خفر
Echinochloa crusgalli	Japanese millet	millet japonais	Japanische Hirse		
Eleusine coracana	finger millet	millet	Korakan. Ragi-Hirse		تيلبون
Hordeum vulgare	barley	orge	Gerste	cebada	شعير
Oryza sativa	rice	riz	Reis	arroz	رز
Panicum miliaceum	millet	millet	Echte Hirse	mijo	
	proso millet				
Pennisetum typhoides	pearl millet	millet perle	Dochan	mijo perla	دُخن
	bullrush millet	millet à chandelle			
Secale cereale	rye	seigle	Roggen	centeno	بركه
Setaria italica	foxtail millet	millet des oiseaux	Kolben-Hirse		
			Welscher		
		millet d'Italie	Fennich		
Sorghum bicolor	sorghum	sorgho	Durra	sorgo	ذُرَة
	guinea corn				
Triticum aestivum	wheat	blé. froment	Weizen	trigo	قمح
Zea mays	maize	mais	Mais	maiz	ذرة شاى
	corn				
PULSES					
Arachis hypogaea	groundnut	arachide	Erdnuss	cacahuete	فول سوداني
	monkeynut	cacahouette		mani	
	peanut				
Cajanus cajan	pigeon pea	pois d'Angole	Taubenerbse	guandu	عدس سوداني
	red gram				
Cicer arietinum	chickpea	pois-chiche	Kichererbse	garbanzo	حمص
	gram				
Dolichos lablab	hyacinth bean	dolique lablab	Lablab-Bohne		لوبيا عفن
	bonavist				

Latin	*English*	*French*	*German*	*Spanish*	*Arabic*
PULSES (contd)					
Glycine max	soya	soja	Soja	soja	
Lens culinaris	lentil	Lentille	Linse	lenteja	عدس مصري
Phaseolus angularis	adzuki		Adzuki-Bohne		
Phaseolus aureus	Mung green gram	haricot de la	Mungbohne	frijal mungo	نبا مسي
	green gram	basse Nubie			
Phaseolus coccineus	runner bean	haricot d'Espagne	Feuerbohne	judia de Espano	
			Prunkbohne		
Phaseolus lunatus	Lima bean	haricot de Lima	Mondbohne	judia Lima	
	butter bean	pois de Cap			
	navy bean				
Phaseolus mungo	black gram	haricot Mungo	Linsen-Bohne	mungo	
Phaseolus vulgaris	haricot bean	haricot	Gartenbohne	judia	فاصوليا
	french bean		Welschebohne		
	snap bean		Buschbohne		
	kidney bean		Gemürebohne		
	wax bean				
	string bean				
Pisum sativum	pea	pois	Erbse	guisante	بسلة
			Ackerbohne		
Vicia faba	broad bean	fève	Feldbohne	haba	فول مصري
	horse bean	feverol	Dickebohne		
	tick bean		Saubohne		
	Windsor bean		Grossbohne		
			Puffbohne		
Vigna unguiculata	cow pea	pois due Bresil	Catjangbohne	caupi	لوبيا علف
	black-eyed pea	dolique du Chine			
		Niebé			

Latin	English	French	German	Spanish	Arabic
HERBAGE & FORAGE CROPS					
Agrostis species	bent fiorin	agrostide	Straussgras Winhalm	agrostide	
Cenchrus ciliaris	buffelgrass African foxtail	cenchrus cilié			
Dactylis glomerata	cocksfoot orchardgrass	dactyle pelotonné	Knaulgras	dactilo apelotonado	
Festuca species	fescue	fétuque	Swingelgras	festuca	
Lolium multiflorum	Italian ryegrass	ray-grass d'Italie	Welsches Weidelgras	raigras annual	
Lolium perenne	perrennial ryegrass	ray-grass anglais	Deutches Weidelgras	raigras perenne	
Medicago sativa	lucerne alfalfa	luzerne	Blaue-Luzerne	alfalfa	برسيم حجازي
Melilotus species	sweet clover	melilot	Steinklee	meliloto	نفل
Paspalum dilatatum	Dallis grass	herbe de Dallis	Brasilianische Futterhirse		
Phleum pratense	Timothy	fléole des prés	Wiesenliesch-gras	colo de topo fleo	
Phalaris stenoptera	Harding grass				
Poa species	meadow grass blue grass	pâturin	Rispengras	poa	
Stylosanthes humulis	Townsville stylo				
Trifolium species	clover	trèfle	Klee	trebol	برسيم

Latin	*English*	*French*	*German*	*Spanish*	*Arabic*
ROOT, OIL & FIBRE CROPS, VEGETABLES & MISCELLANEOUS					
Allium cepa	onion	oignon	Zwiebel	cebolla	بصل
Beta vulgaris	beet	betterave	Rube	remolacha	سلك
Brassica juncea	brown mustard	moutarde brune	Sarepta-Senf		
Brassica napobrassica	swede	rutabaga	Wruke	colinabo	
		navet de Suède	Kohlrube	nabo sueco	
Brassica napus	rape	colza	Raps	nabo	
		navette	Rubsen	colza	
Brassica nigra	black mustard	moutarde noire	Schwarzer Senf	mostaza nigra	خردل أسود
Brassica oleracea	cabbage	chou	Kohl	col-berza	قرنبيط
Brassica pekinensis	Chinese cabbage	chou de Chine	Chinesischer Kohl	col Chino	
Brassica rapa	turnip	navet	Stoppelrube	nabo-colza	لفت
			Steckrube		
Cannabis sativa	hemp	chanvre	Hanf	cáñamo	
Corchorus species	jute	jute	Jute	yute	
Daucus carota	carrot	carotte	möhre	zanahoria	جزر
Dioscorea species	yam		Jamwurzel	batata	
Gossypium species	cotton	coton	Baumwolle	algodón	قطن
Helianthus annuus	sunflower	tournesol	Sommer-Sonnenblume	girasol	عين الشمس
				maravilla	
Hibiscus cannabinus	kenaf	chanvre de Guinée		kenaf	
Lactuca sativa	lettuce	laitue	Garten-Salat	lechuga	خس
			Kopfsalat		
Lycopersicon lycopersicum	tomato	tomate	Tomate	tomate	طماطم
Linum usitatissimum	flax	lin	Lein	lino	
	linseed		Flachs		
Musa species	banana	bananier	Banane	banana	موز
Ricinus communis	castor	ricin	Rizinus	ricino	
Sesamum indicum	sesame	sésame	Sesame	sésamo	سمسم
Sinapis alba	white mustard	moutarde blanche	Weisser Senf	mostaza	خردل أبيض
Solanum tuberosum	potato	pomme de terre	Kartoffel	patata	

Latin	English	French	German	Spanish	Arabic
WEEDS					
Alopecurus myosuroides	black grass	vulpion des champs	Acker- Fuchsschanz		
Avena fatua *A. ludoviciana* }	wild oat	avoine sauvage	Wildhafer Flughafer	avena silvestre	
Cerastium vulgatum	mouse-ear chickweed		Gemeines Hornkraut	pamplina	
Colonyction muricatum	purple moon flower				
Cuscuta species	dodder	cuscute	Fluchsseide Teufelszwirn	cuscuta	
Cyperaceae	sedge	laiche souchet	Riedgras Schilfe Segge	junco	سـبد
Geranium species	cranesbill	bec-de-grue	Storchschnabel	pico de ciguena	
Orobanche species	broomrape	orobanche	Schuppenwurz Zahnkraut		حالوك
Oryza rufipogon	wild red rice				
Ranunculus arvensis	corn crowfoot	renoncule des champs	Acker- Hahnenfus	ranunculo	
Rumex species	docks	rumex	Ampfer	acedera	
Striga hermonthica	witchweed				بودة

GLOSSARY

abnormal seedlings Seedlings which in a germination test do not show the capacity for continued development into normal plants. They may be damaged, deformed, decayed or show other defects.

absorption In this context, the uptake of water molecules into a seed from the ambient air.

after-ripening A physical and/or chemical change which occurs in a ripe seed and results in the termination of dormancy.

amino acids Soluble substances made up basically of carbon, hydrogen and nitrogen atoms, which when combined form proteins.

analytical purity The proportion by weight of a seed lot which comprises intact seeds of the correct species as determined by a test carried out in accordance with the International Rules for Seed Testing.

annual A plant which completes its life cycle from germination to flowering and seed ripening, and then dies within a year.

anther The terminal part of a stamen in which pollen is produced.

anthesis The release of pollen from the anthers.

anthocyanin A red or purple soluble colouring matter in plant tissues.

apomixis Development of a seed without pollination.

auricles The two lobes at the base of the leaf blade which clasp the stem in certain cereals and grasses.

awn A bristle-like prolongation of the glume or lemma in cereals and grasses.

axil The angle between a leaf and the stem which bears it.

basic seed Seed produced by the breeder of a cultivar and intended for the production of certified seed.

biennial A plant which grows vegetatively in the year of sowing; it flowers, sets seed, and then dies in the following year.

bract A small leaf-like structure in an inflorescence, sometimes having a protective function.

breeder's seed A small quantity of seed selected by the breeder of a cultivar for multiplication to produce basic seed.

carbohydrates Complex compounds of carbon, hydrogen and oxygen with the hydrogen and oxygen atoms in the ratio 2:1; present in plants mainly as cellulose, starch and sugars.

caryopsis A one-seeded fruit in which the fruit wall is fused to the seed; typical of cereals and grasses (plural, caryopses).

cereals Species of the grass family cultivated for their edible caryopses.

certified seed Seed of a prescribed standard of quality produced under a controlled multiplication scheme either from basic seed or from a previous generation of certified seed. It is intended either for the production of a further generation of certified seed or for sowing to produce food, forage, etc.

chromosomes Thread-like components of the cell nucleus which carry the genes. The number of chromosomes is the same in all cells and is normally typical of the species, but the number is halved in pollen grains and egg cells.

clone A vegetatively propagated cultivar.

cob The rachis or axis of a maize inflorescence.

coleoptile The sheath surrounding the plumule of the embryo and seedling in the grass family.

cotyledon The leaf of an embryo functioning as a storage or absorptive organ; one or two present in the seed before germination.

combine-harvester A machine which cuts the crop and threshes it in one operation.

commercial seed Seed which is intended for crop production, but has not been produced under a recognized certification scheme.

composite sample A sample obtained by mixing together the primary samples drawn from a seed lot for testing purposes.

corolla Coloured or white flower petals, sometimes fused together to form a tube.

crib A structure with a roof and open sides for drying maize ears.

cultivar A clearly distinguishable group of cultivated plants which, when reproduced under control, retains its distinguishing characters. Equivalent to "variety".

defoliant A chemical which when applied to a plant causes the leaves to fall off, leaving the fruits attached to the stem.

desiccant A drying agent. A field desiccant is a chemical applied to a crop to dry the plant tissues by inhibiting uptake of further moisture through the roots, or by killing the green leaves.

determinate A habit of growth in which the terminal growing point produces an inflorescence or flower, and any further growth of the plant develops from lateral buds.

dormancy The condition in which a seed with a viable embryo fails to germinate in conditions conducive to plant growth.

ear The inflorescence of a cereal plant.

embryo The miniature plant within a seed.

endosperm The nutritive tissue within a seed, but external to the embryo, on which the developing seedling can draw until it is able to photosynthesize on exposure to light.

enzyme A complex substance present in minute quantities, which promotes a biochemical reaction without being changed itself.

epicotyl That part of the seedling stem immediately above the cotyledons and below the first true leaf.

epigeal Type of germination in which the cotyledons are raised above soil level by elongation of the hypocotyl.

evaporation The escape of water molecules (e.g. from a seed) into the ambient air.

F1 hybrid First filial generation arising from a cross between two genetically different parents.

fatty acids Soluble compounds of carbon, hydrogen and oxygen which combine to form fats or oils.

flag leaf The uppermost leaf of a cereal plant with a terminal inflorescence.

flail A stick used traditionally in threshing to separate the seeds from the plant material by beating.

floret A grass flower, or caryopsis enclosed between its lemma and palea. This structure is the "seed" of certain members of the grass family, e.g. species of *Lolium*.

fresh seeds Seeds which absorb water in a germination test but do not germinate; a form of dormancy.

gamete A reproductive cell with the half number of chromosomes, either male or female, which fuses with a corresponding cell of the opposite sex at fertilization.

gene A unit of inheritance which is situated at a fixed point on a chromosome and determines or modifies an inherited character of the plant.

genetic Hereditary.

germinating-ripe The condition of a fully developed seed which is capable of germinating when placed in conditions conducive to plant growth.

germination Resumption of active growth by the embryo. In a test it is regarded as the emergence and development from the seed of those essential structures which indicate the ability of the embryo to develop into a normal plant under favourable field conditions.

germination capacity The percentage of pure seeds which germinate in a standard test to give normal seedlings as defined in the International Rules for Seed Testing.

glumes A pair of chaffy or horny bracts at the base of the spikelet in an inflorescence of the grass family.

grading The process of removing from a seed lot seeds of the same species which are defective, e.g. broken, cracked or discoloured.

grain The caryopsis of a cereal plant, naked or enclosed between a lemma and palea.

hard seeds Seeds which are viable, but fail to absorb water in a germination test or in moist soil; a form of dormancy.

heterozygous The condition in which the two genes in corresponding positions on a pair of chromosomes are different, so that the plant is not true-breeding for the character controlled by these genes.

homozygous The condition in which the two genes in corresponding positions on a pair of chromosomes are the same, so that the plant is pure-breeding for the character controlled by that gene.

hulling The process of removing the tightly enclosing outer covering of a seed, e.g. removing the fruit wall from a clover seed.

husks The leaf sheaths enclosing the ear of maize.

hypocotyl That part of the seedling stem between the cotyledons and the root.

hypogeal Type of germination in which the cotyledons remain below soil level and the plumule is carried upwards by elongation of the epicotyl.

imbibition The initial absorption of water by a seed prior to germination.

inbred Self-fertilized over several generations.

indeterminate A habit of growth in which flowers develop from lateral buds and the terminal growing point continues vegetative growth indefinitely.

inflorescence A group of flowers produced at the apex of a plant or on a lateral branch.

isolation The separation of a seed crop from other crops in order to prevent contamination of the seed to be harvested.

kernel A cereal caryopsis; used particularly with reference to maize.

leaching The removal of soluble substances by soaking in water.

lemma The outer (i.e. on the side opposite the axis) of the two chaffy structures enclosing the flower of a grass plant.

lot See *seed lot*.

maintenance breeder The person or organization indicated in a national cultivar list as responsible for the maintenance of a cultivar and for the production of pre-basic and basic seed of that cultivar; normally the original breeder of the cultivar or his agent.

mass selection The process of improving a population by harvesting seed from superior plants in each generation and bulking them together for sowing to produce the next generation.

meiosis Cell division which produces male or female cells, each with half the full number of chromosomes.

micro-organism A small living organism which can be seen only by means of a microscope, e.g. bacteria.

mitosis Normal cell division in the growing points of plants which produces cells with the same number of chromosomes as the mother cell.

moisture content The weight of water in a seed lot or sample, expressed as a percentage of the total weight of the seed at the time of determination.

mutation A sudden change in a gene, causing a change in the character which it controls.

necrosis Death of part of a plant.

node The point on a stem at which one or more leaves arise.

normal seedlings Seedlings which in a germination test show the capacity for continued development into normal plants when grown in good field conditions.

noxious weeds Weed species which are particularly difficult to eradicate and which are liable to be introduced on to a farm as an impurity in crop seeds; may be designated as such in seed laws.

nucleus The central dense region of a cell which breaks up into chromosomes prior to cell division.

nurse crop A rapidly-growing non-smothering crop, sown and harvested in the first season of a perennial seed crop.

off-type A plant in a seed crop which deviates from the norm for the cultivar, but does not obviously belong to another cultivar or species.

ovary The central organ of the flower which contains one or more ovules and develops into a fruit containing one or more seeds.

ovule The female cell and its associated cells within the ovary which, after fertilization by a male cell, develops into a seed.

palea The inner (i.e. towards the spikelet axis) of the two chaffy structures enclosing the flower of a grass plant.

panicle A branching inflorescence.

paraquat A weedkiller which is lethal to all green tissue—and to man.

parasite An organism which is incapable of photosynthesis and draws nutrients from the living tissues of other plants.

parental material The small stock of a cultivar which the breeder maintains true to type from year to year, and from which he draws plants or seeds to be multiplied to the basic seed stage.

pathogen A micro-organism which lives on or in a living plant, causing a disease.

pellet A single seed encased in a coating of inert matter to form a sowing unit of convenient size.

perennial A plant which grows for more than two years and can flower and set seed in each year, but typically not in the year of sowing.

pericarp The wall of an ovary or fruit.

permeable Allowing water and substances in solution to pass through.

pesticide A substance toxic to insects or micro-organisms attacking plants or plant material, and which is used to control them.

petals The parts of a flower which are white or coloured and attractive to insects.

photosynthesis The formation of carbohydrates in the green parts of a plant under the influence of light.

phytotoxic Poisonous to plants.

picking (maize) Harvesting the whole ears from the plants.

plumule The terminal bud of an embryo or seedling.

pod A non-fleshy fruit containing more than one seed, which dries on maturity and splits open, releasing its seeds; typical of the family Leguminosae.

pollen Yellow grains produced in the anthers of the flower; each grain gives rise to a male nucleus capable of fusing with a female nucleus.

pollination The transfer of pollen grains from an anther to a stigma—followed by fertilization of the ovule.

post-control The process of taking a sample from a certified lot of seed, sowing it in a control plot the following year, and examining the plants to confirm cultivar purity.

pre-basic seed Seed from which basic seed is produced.

pre-control The process of taking a sample from a seed lot which is to be multiplied under a certification scheme, sowing it in a control plot and examining the plants, concurrently with the farm multiplication crops growing from that lot.

primary sample A small portion of seed taken from one point in a seed lot during the sampling process.

proboscis The elongated mouth structure of an insect.

protein Complex substance built up from amino acids.

pubescence Hairs.

pulses Species of the family Leguminosae producing large edible seeds with a relatively high protein content.

pure line A genetically pure population of plants descended from an ancestral single plant by self-pollination in every generation.

pure live seed The percentage by weight of a seed lot consisting of seeds of the named species which are capable of germinating to produce normal seedlings.

quadrat One of the small sample areas of a seed crop which are examined in detail to determine the cultivar purity of the crop as a whole.

quarantine The control of imports for the purpose of preventing the introduction of exotic diseases and pests.

rachilla The branch within a grass spikelet to which the florets are attached; on ripening it may break up leaving a short piece attached to each individual floret.

rachis The axis of an unbranched inflorescence to which the flowers (or. in the grass family, the spikelets) are attached direct.

radicle The rudimentary root of an embryo.

relative humidity The ratio, expressed as a percentage, of the quantity of water vapour actually present in the air to the greatest amount it could contain at that temperature.

Rhizobium A bacterium capable of fixing atmospheric nitrogen, which invades the roots of leguminous plants, stimulating the production of nodule-like swellings.

rogue A contaminant (cultivar, other species or weed) in a seed crop. Roguing is the process of removal from the crop.

royalty Payment made to the owner of a controlling right for the privilege of using the subject (process, design, text or cultivar) so controlled.

saprophyte An organism which is incapable of photosynthesis and feeds on dead plant or animal tissues, e.g. many fungi and bacteria.

scarification Abrasion of the surface of seeds to overcome hardseededness and permit imbibition.

scutellum That part of an embryo in the grass family which corresponds to a cotyledon and absorbs food material from the endosperm.

seed The ripened ovule, or other structure including an ovule, used by farmers as planting material.

seed coat The protective outer layers of a seed.

seedling A young plant from its emergence from the seed until it is established physically and physiologically as a completely independent plant.

seed lot A quantity of seed of one cultivar, of known origin and history, and controlled under one reference number.

semi-permeable Allowing water to pass through, but not dissolved substances.

shelling The process of separating the fruit wall from its enclosed seeds, e.g. in pulses, *or* the separation of maize caryopses from the cob to which they are attached.

silk The long style and stigma of the ovary of maize.

sizing The process of removing seeds smaller and larger than the required dimensions in order to produce a lot of uniform size.

spikelet The unit in an inflorescence of the grass family; consists of one or more florets with a pair of glumes at the base.

stamens The parts of the flower which carry the anthers.

stigma The surface to which pollen grains are transferred for fertilization of ovules (plural, stigmata).

style A long or short extension of the ovary bearing the stigmatic surface.

submitted sample A representative sample taken from a seed lot and submitted to a laboratory for test.

synthetic cultivar A mixture of different interbreeding genetic lines.

tape A collapsed tube of soluble plastic material containing seeds spaced at regular intervals for planting.

tassel The inflorescence of male flowers produced at the apex of a maize plant.

testa The outer covering layer of a true seed, i.e. of a ripened ovule.

thresh Separate the seeds from stems and other plant material.

tiller A branch arising at the base of a cereal or grass plant.

tolerance The maximum acceptable difference between the results of two tests of samples taken from the same seed lot, or between a test result and the standard with which the material is expected to comply.

variety Cultivar.

vector A carrier; usually refers to an insect which transfers pollen from one flower to another, or which carries disease organisms.

weed An unwanted plant appearing accidentally in a cultivated crop.

windrow Cut or uprooted plants gathered loosely into a long row to facilitate drying and subsequent collection.

working sample The sample taken in a laboratory from a submitted sample and actually used in a test.

Index

Index

O

P

Q

R

S

Y

Z